Linked Innovation

Josemaria Siota

Linked Innovation

Commercializing Discoveries at Research Centers

Josemaria Siota
IESE Business School
University of Navarra
Barcelona, Spain

ISBN 978-3-319-60545-6   ISBN 978-3-319-60546-3   (eBook)
DOI 10.1007/978-3-319-60546-3

Library of Congress Control Number: 2017944570

© The Editor(s) (if applicable) and The Author(s) 2018
This work is subject to copyright. All rights are solely and exclusively licensed by the Publisher, whether the whole or part of the material is concerned, specifically the rights of translation, reprinting, reuse of illustrations, recitation, broadcasting, reproduction on microfilms or in any other physical way, and transmission or information storage and retrieval, electronic adaptation, computer software, or by similar or dissimilar methodology now known or hereafter developed.
The use of general descriptive names, registered names, trademarks, service marks, etc. in this publication does not imply, even in the absence of a specific statement, that such names are exempt from the relevant protective laws and regulations and therefore free for general use.
The publisher, the authors and the editors are safe to assume that the advice and information in this book are believed to be true and accurate at the date of publication. Neither the publisher nor the authors or the editors give a warranty, express or implied, with respect to the material contained herein or for any errors or omissions that may have been made. The publisher remains neutral with regard to jurisdictional claims in published maps and institutional affiliations.

Cover Illustration: © saulgranda/Getty

Printed on acid-free paper

This Palgrave Macmillan imprint is published by Springer Nature
The registered company is Springer International Publishing AG
The registered company address is: Gewerbestrasse 11, 6330 Cham, Switzerland

To all managing directors and academic leaders at research centers who positively impact societies and economies through their work

Preface

Why do research centers so often fail to commercialize discoveries? Chap. 1 of this book introduces the core challenge faced by research center managing directors: how to achieve economic sustainability while preserving academic quality. It is a challenging environment, with governments looking to cut costs on R & D, companies not recovering precrisis R & D budgets, investors unwilling to take on such long-term bids, and research centers closing due to the scarcity of resources. It is a paradox that is drawing the attention of leaders in research centers at the university, industry, and government level. It is a growing conversation among academics trying to address the question.

This book covers 28 mechanisms and 12 business models to drive growth in research centers through the examination of performance metrics, design thinking, business modeling, and innovation ecosystems. These mechanisms are followed by successful research centers at universities (e.g., Harvard, Oxford, and Tel Aviv), in industry (e.g., Roche, Apple, and J.P. Morgan), and in government (e.g., NASA, Fraunhofer, and the Chinese Academy of Science). The results are based on an analysis of 3,881 research centers in 107 countries; 54 on-site visits to centers; interviews with 61 leaders—managing directors or those in other leadership roles; a review of 327 publications; and an examination of public data.

Promising to drive economic sustainability in research centers while preserving research quality, linked innovation is framed in Chap. 2 as the connected process between research and commercialization, a route in which the investigation done is transformed into economic value to make the process sustainable. It is a track that interconnects two aspects: the pull of market needs and the push of knowledge. Perceived demand will be met only if the appropriate knowledge is available, and innovation will happen only if there is a market for it. Eight symptoms are then described for identifying the gaps in the innovation process, called broken innovation.

Chapter 3 identifies the four causes behind the failure to select the appropriate research initiatives in early stages of the innovation process: choosing nonholistic performance metrics to decide among projects, a lack of knowledge sharing among agents of the research center, and a lack of either academic or business experience in senior roles. Then, the author examines four practical tools that leading institutions are implementing to solve those problems at research centers: prioritizing projects based on a collection of academic, economic, and social impact metrics, mapping each researcher's focus of study through a research map, and more.

Chapter 4 detects the four causes behind the failure to translate discoveries into inventions: ignorance regarding market need; researchers' lack of business knowledge and engagement with the industry; a scarcity of academic or executive profiles within a research team; and uncoachable researchers. Five hands-on mechanisms are then presented such as translating and mapping consumer needs through design thinking; and following lean research principles by maximizing learning speed and minimizing testing costs.

With regard to the failure to achieve industry collaborations, Chap. 5 uncovers seven causes: an unclear business model; a center's lack of brand; a lack of experience in the research team; an unclear value proposition; a disproportionate research team size; a center's internal bureaucracy or politics; and the unacceptance of research results by external stakeholders. Later, 12 business models being applied at high-performing research centers are presented, including technology transfer through public funding, freemium, consultancy joint ventures, transfer pricing, spin-offs, and so on.

Finally, 10 practical mechanisms are provided for optimizing university-industry-government models and facing the aforementioned barriers: designing a collaborative business model that fits the center's orientation and age; reviewing the processes of your communication unit, ensuring a map of roles, processes in cascade and a CRM of specialized media; and doing periodic industry lectures to translate research results into impact.

Chapter 6 pinpoints the five causes behind not leveraging appropriately a research center's innovation ecosystem. These are a lack of understanding of the innovation ecosystem, internal gaps, no external proximity, a lack of internal resources or hooks to keep talent, and few interactions among the ecosystem's agents. Nine practical mechanisms being applied by recognized centers to confront these issues are then presented, comprising moving from academics to entrepreneurial academics, qualifying the stakeholders of your innovation ecosystem, capitalizing on aging, and so forth.

Answering the initial question of the book, Chap. 7 synthesizes the principles of how to become an economically sustainable research center while preserving research quality, advancing from broken to linked innovation.

Barcelona, Spain   Josemaria Siota

1

Acknowledgements

I would like to thank all those who generously shared their time and provided me with the necessary support to write this book, which could never have been written without their help and contributions.

First, my sincere gratitude goes to Prof. Christoph Zott, Prof. Julia Prats, Prof. Thomas Klueter and Marc Sosna from IESE Business School, who gave feedback on an early proposal and who challenged some of my assumptions. Also to the IESE Ph.D. candidates Melina Moleskis, Susanne Koster, Ali Samei, Stephen Smulowitz and Swapnil Barmase, who provided insightful comments on early drafts. Likewise, to CRUMIC director Alfonso Gironza, who shared observations on a late stage draft. All remaining mistakes are my own.

Second, I would like to give thanks to several experts who provided valuable insights, such as Thatcher Bell, investor in residence at Cornell Tech; Leon Sandler, managing director at the MIT Deshpande Center; Sarah Jane, acting director of MIT's regional entrepreneurship program and manager at Harvard Business School's Institute for Strategy and Competitiveness; Thomas Eisenmann, faculty cochair of Harvard Business School's Rock Center for Entrepreneurship; Jörn Erselius, managing director at Max Planck Innovation; Sam

Breen, former business strategist at the Martin Trust Center for MIT Entrepreneurship; Francesc Subirada, former deputy director at the Barcelona Supercomputing Center; Anjali Sastri, senior lecturer in system dynamics at Massachusetts Institute of Technology; Núria Castell, dean of the Polytechnic University of Catalonia; Marta Ribeiro, former manager of IESE Business School's Center for Research in Healthcare Innovation Management; Alison Baldyga, community manager at the Harvard Innovation Lab; Esther Jiménez, former manager of IESE Business School's International Center for Work and Family; Sidhanth Kamath, former program manager at Ethiopia's Agricultural Transformation Agency; Arnau Tortajada, researcher at IESE Business School's Initiative for Excellence in Operations; Karolina Korth, digital health scout at Roche; Fernando de Sisternes, research affiliate at MIT; Paula Sancho, managing director at IESE Business School's Entrepreneurship and Innovation Center; Carlos Paladella, director of clinical solutions for Southern Europe and Latin America at Elsevier; and Dietmar Tourbier, technology leader for electrical systems at General Electric Global Research.

Third, my gratitude also goes to other contributors, who helped in different endeavors, such as Javier G. Subils, Gabriel Antoja, Christopher Maynard, Pau Amigó, Alessandro Panerai, Víctor Cegarra, Joan Carles Piñol, Miguel Sánchez del Moral, David González, Pau Torné, Timo Sohl, Luca Xianran Lin, Fernando Villar, Nana Yaa Antwi-Gyamfi, Ricardo Benito, Vicent Peris, Giovanni Trebbi, and Cintra Wharton Scott. In addition, my sincere thanks go to numerous people who inspired me every day with their exemplary work, such as Prof. Miguel Ángel Ariño, Álvaro Escrivá, and Jaume Armengou, to name just a few.

Finally, to my family, who was always ready to support me throughout the project.

Contents

1 The Dilemma: Academic Quality or Economic Sustainability — 1

2 From Broken to Linked Innovation: The Underlying Concept — 13

3 Stage 1: Research—Selecting Performance Metrics Based on Academic, Economic, and Social Impact — 27

4 Stage 2: Transformation—Translating Discoveries into Impact for the Market Through Design Thinking — 43

5 Stage 3: Commercialization—Designing Collaborative Business Models for University-Industry-Government Relations — 67

6 All Stages: Innovation Ecosystem—Qualifying and
  Leveraging the Internal and External Agents Based
  on Merit   115

7 Conclusions   143

8 Appendix   147

Index   159

List of Figures

Fig. 1.1	Change in the balance of the policy mix for business innovation	6
Fig. 2.1	Innovation funnel: stages, outputs, and main challenges	15
Fig. 2.2	From broken to linked innovation [18]	22
Fig. 2.3	Categorization of research centers by age and orientation	23
Fig. 3.1	Prioritization of research projects based on holistic impact	31
Fig. 3.2	Main benefits of an international advisory board	40
Fig. 4.1	Example of an innovation funnel from discovery to launch of a new molecular entity	53
Fig. 4.2	Comparison of impact factor (in logarithmic scale) with their rejection rates ($n = 570$ journals)	54
Fig. 5.1	R & D operating costs and equipment in universities by source of funding (Norway)	73
Fig. 5.2	The gap between R & D projects' outcome and company impact	76
Fig. 5.3	Data mining-related scientific articles per 1,000 articles	78
Fig. 5.4	Short-, medium-, and long-term external contracting	78
Fig. 5.5	Internal contracting through transfer pricing	79
Fig. 5.6	Freemium services	80
Fig. 5.7	Research licensing	82
Fig. 5.8	Technology transfer by public funding	83

Fig. 5.9	Spin-off creation from a research center (or a researcher who leaves) via external investment	86
Fig. 5.10	The search model	89
Fig. 5.11	The consultancy joint venture	90
Fig. 5.12	Short-term marketing collaboration	91
Fig. 5.13	Long-term marketing collaboration	93
Fig. 5.14	Percentage of the annual cost incurred in the global scholarly communication process, by value chain component	95
Fig. 5.15	Comparison: size of research teams with total funding (*top*) and indexed papers (*bottom*)	101
Fig. 5.16	Aspects that faculty rated as important for granting tenure	103
Fig. 5.17	Differences in difficulty level between public and private funding	106
Fig. 6.1	Innovation ecosystem (agents and relationships): the triple helix model	121
Fig. 6.2	Innovation ecosystem of Tel Aviv	123
Fig. 6.3	Example of a research center's organizational chart and KPIs	129
Fig. 6.4	Use of digital tools by researchers	132
Fig. 6.5	Motivational measures of academic entrepreneurs	139

List of Tables

Table 1.1	Most satisfying and difficult aspects of being a research center managing director	5
Table 2.1	Challenges and symptoms of broken innovation	19
Table 3.1	Stage 1—research: selecting performance metrics (overview)	30
Table 3.2	Template of a research scorecard	34
Table 3.3	Research map of Harvard Business School's entrepreneurial management unit	36
Table 3.4	Research map of the machine intelligence research group at the University of Cambridge's Department of Engineering	37
Table 4.1	Stage 2—transformation: translating discoveries into impact (overview)	47
Table 4.2	Market map template	51
Table 4.3	Linked innovation map: matching of research and market opportunities	51
Table 5.1	Stage 3—commercialization: designing collaborative business models (overview)	71
Table 5.2	Potential benefits from interactions among universities, industry, and government	74

Table 5.3	Advantages and disadvantages of public and private funding models	84
Table 5.4	Collaborative business model recommended for each type of research center	94
Table 6.1	All stages—innovation ecosystems: leveraging internal and external agents (overview)	119
Table 6.2	Quantification (on the map) and qualification (in the list) of research centers in the innovation ecosystem (Catalonia, Spain)	125
Table 6.3	Simplified visualization of the Global Innovation Index (South Africa)	126
Table 7.1	From broken to linked innovation	145
Table 8.1	Motivations to commercialize discoveries	148
Table 8.2	Examples of each mechanism of linked innovation	148
Table 8.3	A case example for each collaborative business model	153
Table 8.4	Correlation analysis between several performance metrics at research centers ($N = 125$)	156

1

The Dilemma: Academic Quality or Economic Sustainability

Abstract Why do research centers so often fail to commercialize discoveries? This chapter introduces the core challenge faced by research center managing directors: how to achieve economic sustainability while preserving academic quality. It is a challenging environment, with governments looking to cut costs on R & D, companies not recovering precrisis R & D budgets, investors unwilling to take on such long-term bids, and research centers closing due to the scarcity of resources. It is a paradox that is drawing the attention of leaders in research centers at the university, industry, and government level. It is a growing conversation among academics trying to address the question.

Keywords Discovery commercialization · Research center · Economic sustainability · Research funding · Academic quality · Knowledge asset · Technology transfer · Value creation · Value appropriation · Innovation

1.1 What Keeps You up at Night?

One sunny afternoon, before catching a flight to Cologne, I sat in my office chair in Barcelona and started a Skype conversation with an executive 5,860 km away. He was on the leadership team of the Deshpande research center at the renowned Massachusetts Institute of Technology (MIT) in Cambridge.

After a few words, I asked him a tough question: what keeps you up at night? In other words: what is your biggest challenge in leading your research center at MIT? The answer was expected: to achieve a research center that is financially sustainable.

This was the most common answer given by leaders of research centers[1] when asked to identify their top challenge, according to interviews with 61 leaders—managing directors or those in other leadership roles[2]—at 35 international research centers,[3] as well as on-site observation at 28 of those organizations.

No single issue today is higher on the agenda of research centers' senior management than solving this question, whether a center operates in engineering, pharmaceuticals or electronics, or is even a business school [1].

1.2 The Paradox: Quality and Sustainability

Over the past 30 years, the biotechnology industry has attracted more than $300 billion in capital. Much of this investment has been based on the belief that biotech could transform health care through emerging spin-offs from research centers.

These emerging firms were believed to not suffer from the bottlenecks of established pharmaceutical giants, to break down the wall between research-oriented and innovation-oriented centers, and to produce a pool of new drugs. It was assumed that these medicines would create value for society and profits for their investors.

Nevertheless, the promise remains just that. Despite the commercial success of research organizations such as Genentech, which became a subsidiary of Roche, and the revenue growth of the industry, most biotechnology organizations make no profit [2].

So is it possible to make research economically sustainable? There is an implicit paradox, with two priorities that are essential and seem to pull in opposite directions: research quality and economic sustainability.

On the one hand, research centers' academic directors prioritize performance metrics based on academic orientation, focusing on preserving the quality of the research produced and often investing expensive resources. On the other hand, research centers' executive directors prioritize performance metrics based on economic orientation, focusing on ensuring economic sustainability and capturing revenues deriving from discoveries. Is there any way to reconcile and align these opposing perspectives?

Experts follow two streams of thought regarding this paradox. The first stream prioritizes academic metrics, perceiving a tension between "the need for industry funding for academic research and the need to preserve academic freedom" [3] or believing that "working with industry can restrict communication among scientists" for publishing [4]. This tension sometimes results in low levels of revenues to sustain the center economically (e.g., by having a reduced number of industry collaborations).

Examples of this first group include those centers that were created by the United States' National Science Foundation years ago and that have disappeared. After 11 years of public funding, the engineering research centers were expected to become self-sustaining. Many of them were dissolved after completing their federal funding cycle because they did not become economically sustainable [5].

The second stream of thought prioritizes economic metrics, assuming that researchers with industrial support are "at least as productive academically as those without such support and are more productive commercially [6]." This assumption sometimes helps people keep in mind the importance of capturing economic value from discoveries (e.g., generating a large number of industry collaborations), but the quality of the research may decline.

Some examples of this second group are the 159 research centers in 33 different countries that recently disappeared from rankings. Some 70% of the vanished centers emphasized economic over academic metrics. In consequence, these centers either decreased the research quality or limited the creation of knowledge assets,[4] creating an unsustainable cycle of value appropriation. These centers were 14% at the university, 41% in industry, and 45% in government.[5]

Moreover, between the two groups described, there is not only an opposition of thought but also a funding gap. If one looks at the total product development cycle, it seems that the early research stage is typically done in university and government research centers. The later implementation and commercialization part is typically done in industry. However, there is a gap in the middle, which very few organizations address. Therefore, this opposition of thought produces not only a knowledge gap but also a funding gap, which greatly slows down the product development cycle until a "valley of death [7–9]."

Furthermore, external factors make it more difficult to raise public and private funding. On the one hand, local governments are looking for cost savings in the area of research and plan to reduce public funding contributions, while international institutions are changing the rules of funding participation, such as the European Union's Horizon 2,020 program. This program is moving from the principle of full-cost funding to the introduction of lump-sum payments to cover general administrative expenses,[6] which may lead to a deterioration in the funding ratio for EU-sponsored projects [7]. On the other hand, corporations have not yet recovered pre-crisis levels of expenditure on research and development in many countries, such as Finland, Japan, and the United States [10].

As a consequence, many research institutions have applied continuous restructuring, oscillations between research and innovation orientations, centralized to decentralized models and an endless reengineering of processes, with few results [11].

The failure of those organizations is not due to a lack of effort or commitment by management but to the continuing assumption that research centers should choose between academic rigor and economic profitability, with no overlap.

In the midst of this dilemma, research center managing directors hold on, trying to deal with both mind-sets. This complex environment creates benefits and challenges for these managing directors, a role with satisfying and challenging aspects (See Table 1.1).

Table 1.1 Most satisfying and difficult aspects of being a research center managing director

Most satisfying	Most difficult
Intellectual stimulation of working in a cutting-edge innovation environment	Insufficient input in the center's budgetary decisions
	Loss of sponsors
Maintaining relationships and collaborating with the university, industry and government	Insufficient time for multiple activities
	Challenge of motivating faculty members to take opportunities to interact with industry
Impact on economy	Mediating between industry and faculty when projects do not go as planned
Opportunity to work on education and outreach	
Exit career path: known by many industry and university professionals	Multi-institutional coordination
	Protecting intellectual property
	Exchanging ideas that may lead to sponsored research projects

Source Adapted by the author from several sources such as Sander, E. *ERC best practices manual* (National Science Foundation Engineering Research Centers Program, 2013) [12]

However, there is no magic formula because, as in any system, performance is the result of the interaction of many different decisions and choices, including the design of processes for managing projects, the allocation of resources, the mechanisms for transformation and commercialization, and so on. However, we can interiorize decision-making principles that help us drive our organization and learn best practices from top players [1].

In summary, the important question is: why do research centers fail to capture economic value from their knowledge assets and how can they improve without damaging research quality?

1.3 Increasing Interest Among Universities, Industry, and Government

Interest in this topic has increased tremendously not only in existing centers but also among new institutions (i.e., at universities and in industry and government) that are not only advancing in their fields but

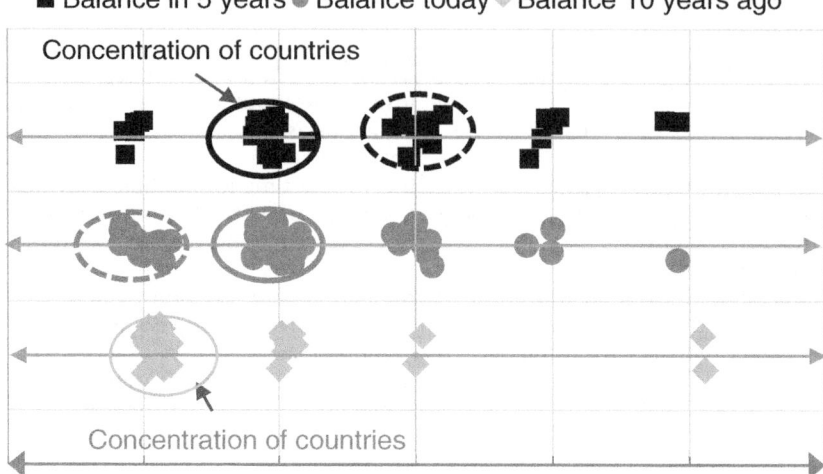

Fig. 1.1 Change in the balance of the policy mix for business innovation. *Source* Adapted by the author from *Science, innovation and technology outlook 2016* (OECD, 2016) [16]. *Note* For simplification and to be consistent with the terminology used in this book, the term "supply-side oriented" was changed to "research oriented" and the term "demand-side oriented" was changed to "innovation oriented"

also using those research centers for new purposes, such as to improve their business models, to create new products, to provide insight to their customers, and so on [13].

In the banking sector, for instance, according to several consultancies (e.g., Accenture), some 20–35% of banking revenues will be at risk in the next 4 years due to the emergence of fintech start-ups [14]. With the risk of losing around one-fourth of their income, the leadership teams at 95% of the analyzed banks have created internal research centers with an orientation toward innovation, during the past years, to discover new business models and new ways to improve the customer journey of their clients.

However, in a recent survey, more than 60% of respondents—chief innovation officers, innovation researchers, user-experience senior consultants and other individuals dealing with innovation—declared their dissatisfaction with the process, not achieving the expected results in either knowledge output or commercial output [15], leaving two questions to be answered: why do research centers fail to capture economic value from their knowledge assets and how can they improve without damaging research quality?

In the government, policymakers have increasingly adopted a broader approach to innovation policy by stimulating demand for innovation, especially in areas of pressing societal need. Many countries indicated that forthcoming years would see increased emphasis on demand-side instruments [16] (Fig 1.1).

1.4 A Growing Conversation Among Academics

In universities, this academic conversation has grown in the past few years—for example, in the *Academy of Management Journal* and the journal *Research Policy* [17], where the number of articles addressing academic-industry engagement and commercialization has increased significantly since 2005, which strengthens the case for a fresh review of the literature [18–24].

The answer is not easy, neither for executives nor for academics—as Professor Pankaj Ghemawat mentioned in a 2015 conference about the connection between academia and business, when he spoke of the "unsatisfactory interchange between academia and the field since the 1990s" [25].

However, there are still many gaps to be filled. For example, in terms of the unit for analysis, an initial inspection of the literature identifies three research streams that examine the elements that enhance collaborative partnerships between academia and industry to improve the capture of value. The first approach is from the university's point of view, [26, 27] the second addresses the issue regarding the firms involved [26, 27] and the third takes the individual academic researcher as the unit of analysis [28, 29].

Second, in terms of the study sample, many articles analyze the United States and selected European countries but contributions covering other geographic contexts are rare [32, 33]. Furthermore, few studies offer cross-national comparative analyses [34–37]. Therefore, the context-specific nature of most published research makes it difficult to reach generalized conclusions.

Third, in terms of available data, many research centers hide their financial information to avoid comparisons and external pressures. Most research so far on academic engagement is phenomenon driven and therefore there are research gaps when it comes to deploying data on this phenomenon to build and test theory [24].

In summary, there was a need to work out how to capture economic value from knowledge assets without damaging research quality at research centers, from the organization's perspective as a unit of analysis, with evidential data of multinational origin.

This book tries to provide answers for research center directors, for professional service firms that are emerging to solve this kind of problem, and for academics who are trying to advance to new solutions.

The results are based on an analysis of 3,881 research centers in 107 countries; 54 on-site visits to centers; 61 interviews with research center managing directors, politicians, consultants, investors, advisers, professors, executives, and researchers; a review of 327 publications; and an examination of public data.

Endnotes

1. "Research center" is understood as being an establishment empowered to do research, which comprises "creative work undertaken on a systematic basis in order to increase the stock of knowledge, including knowledge of humans, culture and society, and the use of this stock of knowledge to devise new applications"—this definition is from the *Frascati Manual* (OECD, 2002). Throughout this manuscript, to keep matters simple, I use the term "research centers" to refer to such centers at universities and in industry and government (such as university departments, corporate research and innovation labs, and public research institutions).

2. Other senior leadership roles at research centers included academic directors and directors of technology transfer offices. Other complementary roles included politicians, consultants, investors, advisers, professors, and researchers.
3. The research centers to which the interviewees were affiliated were located in Germany, the United States, Spain, and China.
4. The innovation outputs of these vanished centers were higher than their knowledge outputs.
5. (1) The analyzed data come from Elsevier's Scopus database. The data of a 6-year period were accessed on 2014 and were later classified. (2) Institutions were selected with the criterion that they needed to be research institutions with more than 100 published works included in the Scopus database. (3) Subinstitutions were counted as one, the parent. As a consequence, each research institution may include more subinstitutions.
6. A lump-sum payment is a single payment made at a particular time, as opposed to a number of smaller payments or installments [38]. Generally, the total value of a lump-sum payment is less than several payments because the payer has to provide the funds in advance.

References

1. Hannon, E., Smits, S. & Weig, F. Brightening the black box of R & D. *McKinsey Quarterly* 1–11 (2015). [Issue 2]
2. Pisano, G. P. Can science be a business? Lessons from biotech. *Harvard Business Review* **10**, 1–10 (2006).
3. Lee Yong, S. University–industry collaboration on technology transfer: Views from the ivory tower. *Policy Studies Journal* **26**, 69–84 (1998).
4. Welsh, R., Glenna, L., Lacy, W. & Biscotti, D. Close enough but not too far: Assessing the effects of university–industry research relationships and the rise of academic capitalism. *Research Policy* **37**, 1854–1864 (2008).
5. Boschi, F. C. *Best Practices for Building and Maintaining University–Industry Research Partnerships: A Case Study of Two National Science Foundation Engineering Research Centers* (Montana State Univ., 2005).
6. Perkmann, M. *et al.* Academic engagement and commercialisation: A review of the literature on university–industry relations. *Research Policy* **42**, 423–442 (2013).

7. Fraunhofer Gesellschaft. *Annual Report 2013: Living in the Digital World.* (2013).
8. Frank, C., Sink, C., Mynatt, L., Rogers, R. & Rappazzo, A. Surviving the 'valley of death': A comparative analysis. *The Journal of Technology Transfer* **21**, 61–69 (1996).
9. Auerswald, P. E. & Branscomb, L. M. Valleys of death and Darwinian seas: Financing the invention to innovation transition in the United States. *The Journal of Technology Transfer* **28**, 227–239 (2003).
10. OECD. *Main Science and Technology Indicators—Industry-Financed Gerd as a Percentage of GDP* (2015).
11. Pisano, G. Creating an R & D strategy. *Harvard Business School Working Knowledge* 1–9 (2012).
12. Sander, E. *ERC Best Practices Manual Chapter 5 Industrial Collaboration and Innovation.* National Science Foundation (2013).
13. Rohrbeck, R. & Arnold, H. M. Making University–industry collaboration work—a case study on the Deutsche Telekom Laboratories contrasted with findings in Literature. In *ISPIM Annual Conference: Networks for Innovation* (2007).
14. Busch, W., Pigliucci, A., Mulhall, R., Rjeily, A. & Wallis, J. *The Digital Disruption in Banking—Demons, Demands, and Dividends* (Accenture, 2014).
15. Siota, J., Klueter, T., Staib, D., Taylor, S. & Ania, I. *Design Thinking: The New DNA of the Finacial Sector* (IESE Business School; Oliver Wyman, 2017).
16. OECD. *Science, Technology and Innovation Outlook 2016* (2016).
17. Pfeffer, J. A modest proposal: How we might change the process and product of managerial research. *Academy of Management Journal* **50**, 1334–1345 (2007).
18. Perkmann, M. & Walsh, K. University–industry relationships and open innovation: Towards a research agenda. *Internal Journal of Management Reviews* **9**, 259–280 (2007).
19. Phan, P. H. & Siegel, Donald S. The effectiveness of university technology transfer. *Foundations and Trends in Entrepreneurship* **2**(2), 77–144 (2006).
20. Rothaermel, F. T., Agung, S. D. & Jiang, L. University entrepreneurship: A taxonomy of the literature. *Industrial and Corporate Change* **16**, 691–791 (2007).
21. Geuna, A. & Muscio, A. The governance of university knowledge transfer. *Sewps* **44**, 14 (2008).

22. Larsen, M. T. The implications of academic enterprise for public science: An overview of the empirical evidence. *Research Policy* **40**, 6–19 (2011).
23. Phan, P. H., Siegel, D. S. & Wright, M. Science parks and incubators: Observations, synthesis and future research. *Journal of Business Venturing* **20**, 165–182 (2005).
24. Perkmann, M. *et al.* Academic engagement and commercialisation: A review of the literature on university–industry relations. *Research Policy* **42**, 423–442 (2013).
25. Ghemawat, P. *Business Strategy Interfaces and Frontiers Conference.* (2015).
26. Di Gregorio, D. & Shane, S. Why do some universities generate more start-ups than others? *Research Policy* **32**, 209–227 (2003).
27. Friedman, J. & Silberman, J. University technology transfer: Do incentives, management, and location matter? *The Journal of Technology Transfer* **28**, 17–30 (2003).
28. Cohen, W. M., Nelson, R. R. & Walsh, J. P. Links and impacts: The influence of public research on industrial R & D. *Management Science* **48**, 1–23 (2002).
29. Fontana, R., Geuna, A. & Matt, M. Factors affecting university–industry R & D projects: The importance of searching, screening and signalling. *Research Policy* **35**, 309–323 (2006).
30. D'Este, P. & Patel, P. University–industry linkages in the UK: What are the factors underlying the variety of interactions with industry? *Research Policy* **36**, 1295–1313 (2007).
31. Berbegal-Mirabent, J., Sánchez García, J. L. & Ribeiro-Soriano, D. E. University–industry partnerships for the provision of R & D services. *Journal of Business Research* **68**, 1407–1413 (2015).
32. Giuliani, E., Morrison, A., Pietrobelli, C. & Rabellotti, R. Who are the researchers that are collaborating with industry? An analysis of the wine sectors in Chile, South Africa and Italy. *Research Policy* **39**, 748–761 (2010).
33. Walsh, J. P., Baba, Y., Goto, A. & Yasaki, Y. Promoting university–industry linkages in Japan: Faculty responses to a changing policy environment 1. *Prometheus* **26**, 39–54 (2008).
34. Dutrénit, G. & Arza, V. Channels and benefits of interactions between public research organisations and industry: Comparing four Latin American countries. *Science and Public Policy* **37**, 541–553 (2010).
35. Grimpe, C. & Fier, H. Informal university technology transfer: A comparison between the United States and Germany. *The Journal of Technology Transfer* **35**, 637–650 (2009).

36. Haeussler, C. & Colyvas, J. A. Breaking the ivory tower: Academic entrepreneurship in the life sciences in UK and Germany. *Research Policy* **40**, 41–54 (2011).
37. Klofsten, M. & Jones-Evans, D. Comparing academic entrepreneurship in Europe—the case of Sweden and Ireland. *Small Business Economics* **14**, 299–309 (2000).
38. Oxford Dictionaries. *Definition of lump-sum.* https://en.oxforddictionaries.com/definition/lump_sum (2017).

2

From Broken to Linked Innovation: The Underlying Concept

Abstract Promising to drive economic sustainability in research centers while preserving research quality, linked innovation is framed in this chapter as the connected process between research and commercialization, a route in which the investigation done is transformed into economic value to make the process sustainable. It is a track that interconnects two aspects: the pull of market needs and the push of knowledge. Perceived demand will be met only if the appropriate knowledge is available, and innovation will happen only if there is a market for it. Eight symptoms are then described for identifying the gaps in the innovation process, called broken innovation, through the examination of four challenges: performance metrics, market understanding, industry collaborations, and innovation ecosystems.

Keywords Broken innovation · Linked innovation · Knowledge push · Market pull · Invention · Innovation funnel · Discovery commercialization · Knowledge transfer · Research center · Economic sustainability · Value creation · Value appropriation

2.1 The Stages of the Innovation Funnel: Research, Transformation, and Commercialization

Last week, I met an executive of the healthcare company Roche. We were discussing a study, published in the journal *Research-Technology Management*, that describes how most industries appear to require 3,000 raw ideas to produce one substantially new, commercially successful industrial product.

3,000 raw ideas turn into 300 ideas for which minimal action is taken—such as simple experiments, patent filing, or management discussion. Then 125 of those 300 ideas become small projects; 9 out of the 125 become significant projects with a significant development effort; 4 out of the 9 become major development efforts; 1.7 out of the 4 is commercially launched; and 1 out of the 1.7 launched (59%) becomes commercially successful. (This last success rate varied from 40 to 67%, depending on the source of information, industry, and geography.) [1].

Joseph A. Schumpeter, who is regarded as one of the greatest economists of the first half of the twentieth century, explained this process of innovation as "the same process of industrial mutation […] that incessantly revolutionizes the economic structure from within, incessantly destroying the old one, incessantly creating a new one" [2].

This mutation process is analogous to the three classic states of matter. First, the research activity generates discoveries that are intangible and highly adaptable, like a gas. Second, the transformation activity transforms discoveries into inventions that are modestly tangible but still adaptable, like a liquid. Third, commercialization activity transforms inventions into innovations that are tangible and nonadaptable, like a solid [3, 4].

Across each activity of the innovation process—research, transformation, and commercialization—there is an innovation output: discovery, invention, and innovation. Although many discoveries start at the research stage, few inventions reach the commercialization stage.

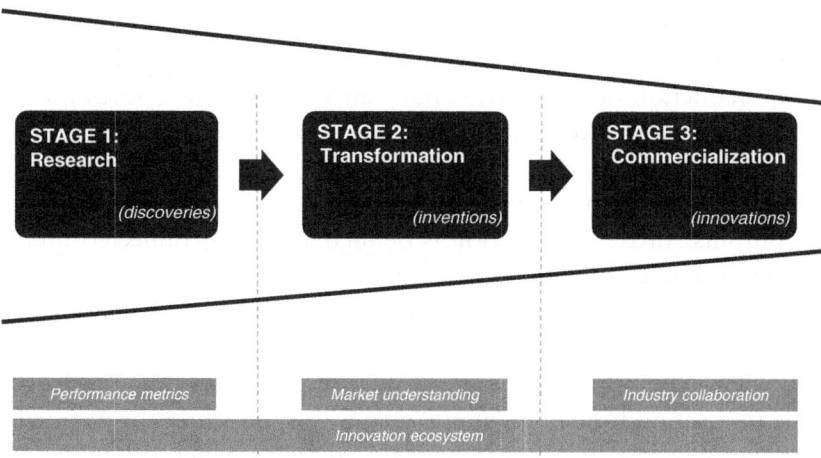

Fig. 2.1 Innovation funnel: stages, outputs, and main challenges. *Source* Prepared by the author. *Note* In each stage, the *top-left corner* shows the activity and the *bottom-right corner* shows the output. In the literature, authors use several names for each stage (e.g., research, development, innovation). However, for simplification, after the interview process, these names were chosen by contrasting the perception of academics and that of practitioners

In summary, we can visualize this selection process as an innovation funnel with three stages. First is the research,[1] which includes the generation of discoveries or knowledge assets. Second there is the transformation, which includes the translation of knowledge assets into tangible assets—inventions. Third is the commercialization, which includes the capturing of economic value, with those tangible assets being transformed into commercial assets—innovations[2] (See Fig. 2.1).

2.2 Commercializing Discoveries: Main Challenge Areas

Worldwide innovation capital represents 42% of gross domestic product (GDP), increasing at a rate of 4.6% a year (as of 2013). Research and development account for 23% of the innovation capital [5]. Although innovation has many benefits, the process has several barriers.

The most important motives for commercializing discoveries at research centers are the extension of the product or market portfolio, the reduction of entry costs into new markets, the amortization of R & D expenditure and the strengthening of technological leadership (See Sect. 8.1) [6].

However, research inputs do not always trigger commercialization outputs. Broken innovation is defined as the unconnected process between research and commercialization, a route in which the investigation undertaken is not transformed into economic value to make the process sustainable.

So what makes this process unconnected? To identify the bottlenecks in the innovation funnel, my initial goal was to prioritize the main challenges faced by research center managing directors when trying to commercialize their internal discoveries along the stages of the innovation funnel.

My results suggest that there are four main challenge areas for capturing value from discoveries: performance metrics, market understanding, industry collaboration, and the innovation ecosystem [6, 7].

First, the performance metrics are used to decide how to choose from among different initiatives in the research stage. In other words, how do I grade, prioritize, and select from several initiatives. For example, should I measure the number of potential new publications? Should I measure the synergies I expect to achieve within the institution? Should I measure the number of people who may be cured of the illness being studied? And so on.

Second, there are principles and actions for understanding the market and successfully transforming discoveries into products and services. For instance, how to choose among different transformations of a discovery—an article, a patent, a license, a spin-off, etc.? How do I know what the market demands?

Third, there are mechanisms to apply, during industry collaborations, to commercialize discoveries effectively. For example, how do I collaborate with the industry, keeping in mind its tight schedules and short-term demands for returns on investment? How do I keep the academic freedom to avoid biases in the process? How do I commercialize the inventions?

Fourth, in the innovation ecosystem, there are ways to fill stakeholder gaps throughout the process.[3] For instance, how do I commercialize discoveries if there are no companies with enough revenue to pay for those products or services? How can you overcome the internal bureaucracy involved in releasing intellectual property when you are a researcher and not a legal expert? How do you launch experiments when you do not have the labs or internal capabilities to do so? (See Fig. 2.1).

In each challenge area, I identified the differences between those centers that achieved a connected process among the innovation stages and those that did not. Is there any way to identify the unconnected process through those patterns?

2.3 Broken Innovation: The Symptoms

1772, 1929, 1973, 1997, and 2007. If you like to play with numbers, what do these five have in common?

These are the approximate starting dates of five of the world's most devastating financial crises. Financial crises are, unfortunately, quite common in history and often cause economic tsunamis.

It was difficult to predict those crises. However, in a lesson learned, policymakers have suggested reducing the risk by identifying symptoms and monitoring vulnerabilities. Possible alarm signals are asset price bubbles and credit booms, which can entail substantial costs if they deflate rapidly. A monitoring of sharp increases, for instance, may help to determine whether they would be followed by a steep and rapid decline [8].

There are still signals that some research centers are disappearing (See Sect. 1.2). Is there any way to identify the symptoms of broken innovation to avoid a Lehman Brothers-style crash among research centers?

In some cases, problems can be predicted by identifying symptoms which are indicators of negative underlying issues that can potentially lead to an undesirable situation or unwanted scenarios (e.g., broken innovation).

My results suggest eight scenarios that trigger broken innovation: two triggers per each identified challenge. These symptoms have been identified through an analysis of research centers, the triangulation of public information, interviews with their managing directors, on-site visits, a review of the available literature, and so on (See Sect. 8.4) [9].

For each problem across the three stages of the innovation funnel, the symptoms are illustrated as well as the two most common scenarios for each case (See Table 2.1). These triggers are a thermometer to identify whether you are suffering because of the undesired situation.

First, when deciding on the performance metrics on how to pick the best one among several initiatives during the research process (stage 1), there were two triggers. There were the centers that prioritized only academic metrics (e.g., the number of refereed articles published in top journals) used to find their economic sustainability was on a downward slope. Alternatively, there were the centers that prioritized only economic metrics (e.g., profits) also experienced a decrease in the quality of their research. In summary, there was a lack of alignment in priorities.

These two symptoms can be expressed in questions for self-reflection: are you experiencing a decline in research quality? This might involve, for example, whether you perceive a decline in the research projects' rigor or in the number and quality of the center's publications, or whether you experienced external biases from collaborators. Are you facing a decrease in economic profitability?

Second, when it came to understanding the market to translate discoveries into products and services during the transformation process (stage 2), there were two triggers. There were the centers that assumed what the market needed rather than validating what the market actually wanted (in terms of research questions, feedback, etc.). Then, there were the centers that were merely faithful followers of market trends, closing the door on creating something beyond those trends. In summary, there was a lack of alignment in needs.

These two symptoms can be expressed in questions for self-reflection: are you coming up with products or services that no one wants to buy? Are you producing outdated products? Are your researchers always changing their focus and research interests?

Third, when it came to managing collaborations with the industry in the commercialization process (stage 3), there were two triggers. On the one hand were the centers that focused on research collaborations

2 From Broken to Linked Innovation: The Underlying Concept

Table 2.1 Challenges and symptoms of broken innovation

	Stage 1: research	Stage 2: transformation	Stage 3: commercialization
Broken innovation	**Performance metrics** Economic vs. academic Are you facing a decline in research quality or in economic profitability?	**Market understanding** Assuming vs. following Are you coming up with products that no one wants to buy or that are outdated?	**Industry collaboration** Research vs. furtive Are you experiencing increased difficulty in monetizing your discoveries or in getting access to industry data?
		Innovation ecosystem Internal vs. external Do you lack an innovation catalyst inside or outside your organization?	

Source Prepared by the author based on analysis of the cases, interviews, and a literature review [10–16]

only to gather data and nurture databases to develop their own projects, without giving much consideration to how to help the other stakeholder. On the other hand were the centers that were seen as furtive sales centers, whose only goal was to capture economic value from research results. In summary, there was a failure to ensure that both stakeholders benefited.

These two symptoms can be expressed in questions for self-reflection: are you experiencing increased difficulty in monetizing your discoveries? Are you experiencing increased difficulty in getting access to industry data and professional networks?

Finally, there was the matter of choosing the right internal and external partners to enhance your center across the three stages of the innovation funnel. There were the centers that chose their partners according to internal proximity—those inside their offices or institutions—and assumed that the whole innovation ecosystem should be kept inside the organizations. Alternatively, there were the centers that prioritized partners according to external proximity—those that were geographically near their research centers—and assumed that the whole innovation ecosystem should neighbor their organizations. In summary, proximity was prioritized rather than meritocracy.

These two symptoms can be expressed in questions for self-reflection: do you lack innovation catalysts inside your organization? In other words, is your research center prevented from successfully commercializing its discoveries by the fact that the center is missing an actor? Otherwise, do you lack innovation catalysts outside your organization?

If you have answered "yes" to any of the previous four pairs of questions, you are probably suffering the symptoms of broken innovation.

2.4 Linked Innovation: Mixing the Knowledge Push and Market Pull

During the past 4 years, several directors of research centers in the private and public sector, who were aware of this research project and who were conscious that their institutions were experiencing a process of broken innovation, came to me and said: "OK, Josemaria, now I know that I have a broken innovation process. How can I solve the problem? How can I solve the lack of a connection between the research quality of my center and the economic results?"

After identifying the four common challenges and symptoms in terms of commercializing discoveries across the innovation funnel in a process of broken innovation (See Table 2.1), the following sections will set out the causes of the problems identified and suggest several solutions for moving to linked innovation.

Linked innovation is defined as the connected process between research and commercialization, a route in which the investigation done is transformed into economic value to make the process sustainable.

This model of linked innovation is based on several models, such as the chain-linked innovation model of Kline and Rosenberg that embodies the technical activities occurring in the innovation process, the external forces of the marketplace, and the complex interactions between the various stages of the process [17].

In my proposed model of linked innovation, two aspects are interconnected: the pull of market needs and the push of knowledge (See Fig. 2.2). In other words, you try to get the best of both the offer side and the demand size. Perceived demand will be met only if the appropriate knowledge or technology is available, and innovation will be realized only if there is a market for it.

This model is a virtuous circle in which the more resources we receive from research, the more resources we can invest in additional research, and so on.

Although the identified problems and solutions are applicable to all types of research centers (e.g., from health care to engineering, and from science to business), the following sections will include a subtle classification using two independent variables that affect the commercialization of discoveries at research centers. These two variables will help you to identify, in more detail, when to focus on a problem or prioritize a solution, depending on the characteristics of your center.

The first segmentation is the age or experience of the center. Research centers that have been created within the past 7 years and that are at an

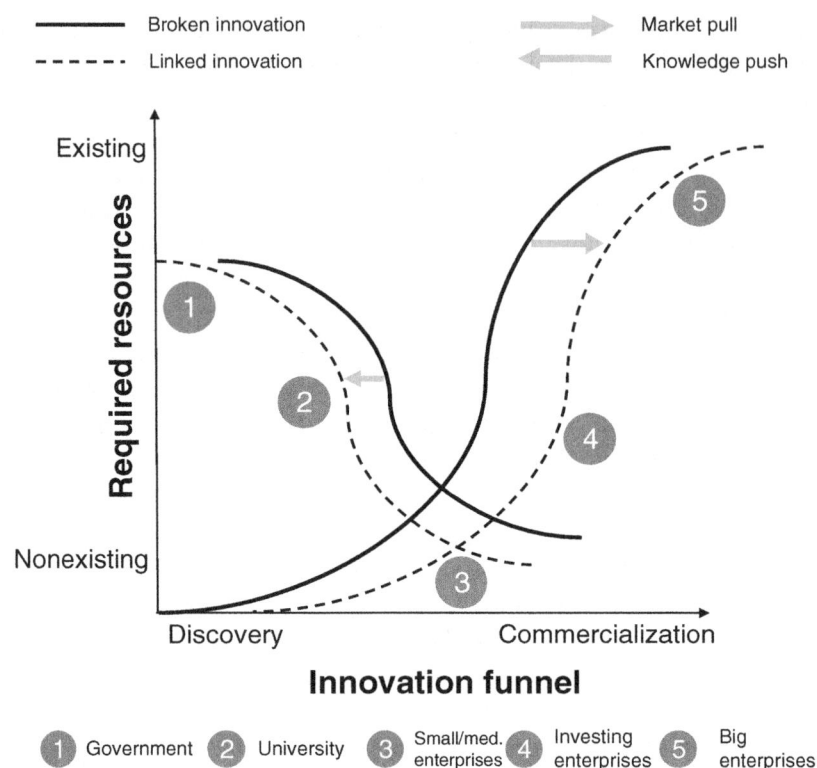

Fig. 2.2 From broken to linked innovation [18]. *Source* Adapted by the author from several models such as *ERC best practices manual* (National Science Foundation Engineering Research Centers Program, 2013)

early stage will be considered young. Otherwise, they will be considered mature. The second segmentation is by orientation. The research center is focused either on research (e.g., answering more theoretical questions, which is usually the case with centers in universities) or on innovation (e.g., answering more practitioner-oriented questions, which is usually the case with centers in industry) [19]. These two variables (age and orientation) give us four possible segments (See Fig. 2.3).

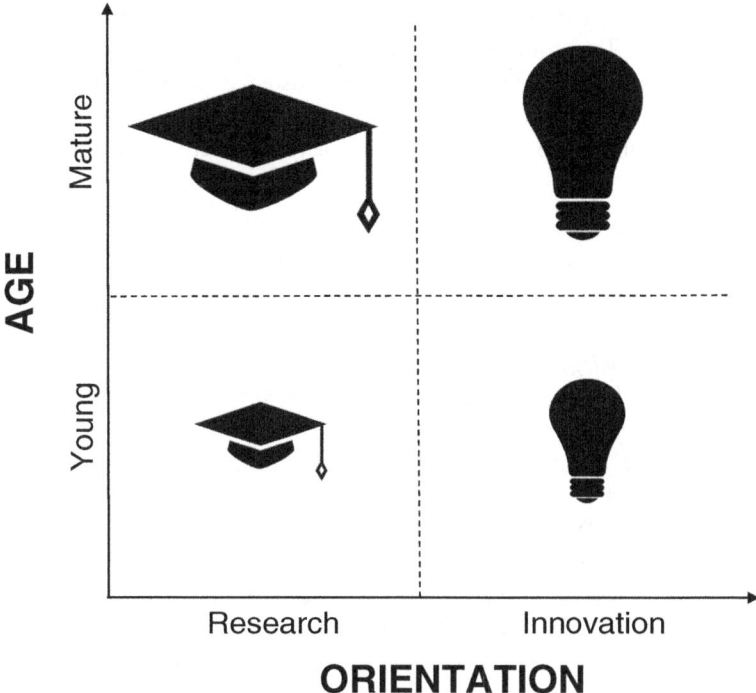

Fig. 2.3 Categorization of research centers by age and orientation. *Source* Prepared by the author

Endnotes

1. According to the Oxford English dictionary, the word "research" means "the systematic investigation into and study of materials and sources in order to establish facts and reach new conclusions" and the word "innovation" means "the action or process of creating a new method, idea, product, etc."
2. This is a simplification of the chain-linked innovation model of S.J. Kline and N. Rosenberg: An overview of innovation in *The positive sum strategy: harnessing technology for economic growth* (ed. Landau, R., and Rosenberg, N.), 275–305 (National Academy Press, Washington, DC, 1986).

3. An innovation ecosystem is as a set of components, relationships and functions. The components are stakeholders who participate in that environment and are necessary for its existence (from the point of view of both supply and demand). The existence of all of these stakeholders is as important as the nature of the multilateral interrelationships.

References

1. Stevens, G. A. & Burley, J. 3000 raw ideas equal 1 commercial success! *Research Technology Management* **40**, 16–27 (1997).
2. Schumpeter, J. A. Creative destruction. *Capitalism, Socialism and Democracy* 82–85 (George Allen and Unwin, 1942).
3. Lane, J. P. & Flagg, J. L. Translating three states of knowledge–discovery, invention, and innovation. *Implementation Science* **5**, 9 (2010).
4. Roberts, E. B. What we've learned: managing invention and innovation. *Research Technology Management* **31**, 11–29 (1988).
5. Baily, M., Haskel, J., Hazan, E., Marston, N. & Rajah, T. *Innovation Matters Reviving The Growth Engine* (McKinsey, 2013).
6. Lichtenthaler, U. External commercialization of knowledge: Review and research agenda. *International Journal of Management Reviews* **7**, 231–255 (2005).
7. Institute of Medicine (US). *Forum on Drug Discovery, development, and translation Challenges in Clinical Research* (2010).
8. Kose, A. & Claessens, S. *Financial Crises Explanations, Types, and Implications* (International Monetary Fund, 2013).
9. Şendoğdu, A. A. & Diken, A. A research on the problems encountered in the collaboration between university and industry. *Procedia—Social and Behavioral Sciences* **99**, 966–975 (2013).
10. Siegel, D. S., Waldman, D. A., Atwater, L. E. & Link, A. N. Commercial knowledge transfers from universities to firms: improving the effectiveness of university–industry collaboration. *The Journal of High Technology Management Research* **14**, 111–133 (2003).
11. Etzkowitz, H. & Leydesdorff, L. The dynamics of innovation: from national systems and "Mode 2" to a triple helix of university-industry-government relations. *Science and Technology* **29**, 109–123 (2000).
12. Azároff, L. V. Industry–University collaboration: How to make it work. *Research Management* **25**, 31–34 (1982).

13. Van Dierdonck, R. & Debackere, K. Academic entrepreneurship at Belgian Universities. *R & D Management* **18**, 341–353 (1988).
14. Cyert, R. M. & Goodman, P. S. Creating effective university–industry alliances: an organizational learning perspective. *Organizational Dynamics* **25**, 45–57 (1997).
15. Faems, D., Van Looy, B. & Debackere, K. Interorganizational collaboration and innovation: toward a portfolio approach. *Journal of Product Innovation Management* **22**, 238–250 (2005).
16. Vohora, A., Wright, M. & Lockett, A. Critical junctures in the development of university high-tech spinout companies. *Research Policy* **33**, 147–175 (2004).
17. Kline, S. J. & Rosenberg, N. *The Positive Sum Strategy* 275–305 (National Academies Press, 1986).
18. Sander, E. *ERC Best Practices Manual Chapter 5 Industrial Collaboration and Innovation* (National Science Foundation, 2013).
19. Perkmann, M. *et al.* Academic engagement and commercialisation: A review of the literature on university–industry relations. *Research Policy* **42**, 423–442 (2013).

3

Stage 1: Research—Selecting Performance Metrics Based on Academic, Economic, and Social Impact

Abstract This chapter identifies the four causes behind the failure to select the appropriate research initiatives in early stages of the innovation process: choosing nonholistic performance metrics to decide among projects, a lack of knowledge sharing among agents of the research center, and a lack of either academic or business experience in senior roles. Then, the author examines four practical tools that leading institutions are implementing to solve those problems at research centers: prioritizing projects based on a collection of academic, economic, and social impact metrics; mapping each researcher's focus of study through a research map and incentivizing collaborations and sharing the best practices among them; using professional recruitment for academic and executive directors; and attracting an international advisory board.

Keywords Performance metrics · Academic rigor · Economic relevance Academic impact · Economic impact · Social impact · Research map Academic recruitment · Knowledge sharing · Academic incentives · Research center · Linked innovation

It was 4 pm in Barcelona and 10 am in Massachusetts. I connected to Skype. On the other side of the Atlantic Ocean was an executive from the MIT Deshpande Center for Technological Innovation. Established in 2002, the center empowers some of MIT's most talented researchers to make a difference in the world by developing innovative technologies in the lab and bringing them to the marketplace in the form of breakthrough products and new companies. In turn, those innovations can help solve large and daunting problems in health, information technology, energy, and other fields.

The executive explained that he and his team did not see a lot of difference between developing a new technology in the lab, starting a company, and engaging in philanthropy. All these activities are about making a difference: you see something you do not like in the world, something you believe can be better, and you resolve to improve it by creating a new technology, a new product, or a new service.

By showing these very talented researchers how to commercialize their innovations so they have a real-world impact, the MIT Deshpande Center is helping them become change makers—people who do not leave things at the idea stage but push ahead and actually make their ideas happen.

After getting an understanding of the center's vision and mission, I asked the executive a tough question: what keeps you up at night? The first challenge is to be economically sustainable. Next? How to prioritize among several research projects.

3.1 From Causes of Broken Innovation to Best Practices for Linked Innovation

Are you experiencing a decline in research quality? Are you experiencing a decrease in economic profitability? These were the two most common symptoms identifying broken innovation during the research stage in the performance metrics. In the first case, you are probably prioritizing only economic metrics in your decision-making process. In the second case, you are probably prioritizing only academic metrics in your decision-making process.

We should understand the core problem at a conceptual level initially in order to solve the problem at an action level. Researchers and administrators have long held different views about how to measure performance and freedom among research centers [1]—a dilemma between two orientations: value creation and value capturing.

Researchers want a greater level of freedom to do research into what they like the most without pushing metrics. They think that greater involvement in pure research projects will give them a better track record in their academic careers through the publication of more papers in high-ranking journals and better citations for their previously published work, earning a name for themselves among the academic community [2].

Nevertheless, administrators need to ensure that the center is economically sustainable so as to be able to hire more researchers, ensure there is a well-prepared ecosystem to support the researchers' activities (e.g., securing funds for their researchers, graduate students and lab equipment, and field-testing the application of the research) [3] and provide outreach and make their results visible [4].

Among the analyzed cases, four causes were identified as a trigger of the symptoms of broken innovation in research (stage 1). These four aspects are project prioritization, interconnectivity, knowledge sharing, and leadership experience.

First, a nonholistic prioritization of the research projects gives too much emphasis to a particular criterion. On the one hand are the quick wins, the returns on risk, and the pace of release of new products. On the other hand, there are criteria that consider only the academic value and rigor, forgetting the relevance to society or to the market.

Second, there is the creation of silos or independent research teams. There are units that do not know what other teams are doing and may be involved in similar projects, so creating duplication and avoiding cost optimizations at a research center level.

Third, there is the director's professional experience. Sometimes directors come from a nonacademic background. In many cases, these roles are selected by the institution's internal community, with the experience in the institution given priority as one of the key factors rather than other criteria being used or external advisers or professional recruiters being leveraged.

Table 3.1 Stage 1—research: selecting performance metrics (overview)

Orientation and age of the center					
Research young	Research mature	Innovation young	Innovation mature	Causes of broken innovation	Best practices for linked innovation
x	x	x		(a) Project: nonholistic prioritization	(a1) Prioritize research projects based on holistic impact
	x		x	(b) Interconnectivity: lack of knowledge sharing	(b1) Research map: map the research interests of each researcher and the research projects of the center
x	x	x		(c) Leadership: lack of nonacademic experience	(c1 and d1) Use professional recruitment for academic and executive directors
x		x	x	(d) Rigor: lack of academic experience	(c2 and d2) Attract and recruit an international advisory board

Source Prepared by the author

Fourth is a lack of academic experience, seen as a shortage of work already published in top academic journals. This scenario may also include unknown biases in the results, overselling assumptions, etc.

These causes were common to a greater or less extent, depending on the center type—its age and orientation. In the following sections, we will look at the four best practices that solve each of the identified problems (See Table 3.1).

3.2 Prioritize Research Projects With Performance Indicators Based on Holistic Impact

Are you experiencing a decline in research quality? Are you not economically sustainable or experiencing a decrease in economic profitability?

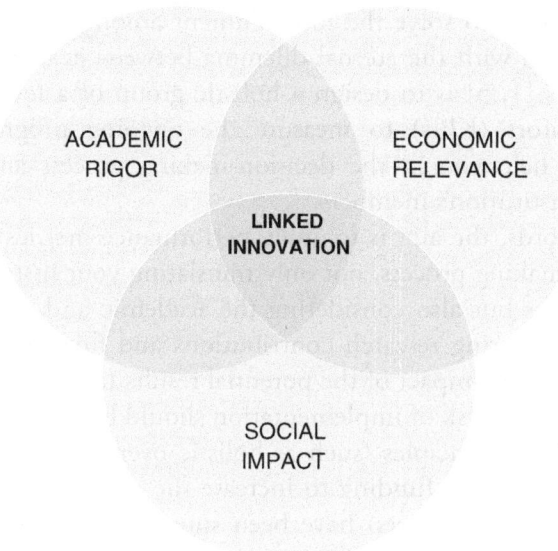

Fig. 3.1 Prioritization of research projects based on holistic impact. *Source* Prepared by the author

The risk in this scenario is not low. Examples of centers that have disappeared provide a learn-by-example lesson. For instance, there are the centers that disappeared because of a lack of economic sustainability after the 11 initial years of public funding offered by the National Science Foundation. There are also the 159 centers that were removed from the rankings for, among other reasons, not keeping quality of research as one of their priorities (See Sect. 1.2).

The institutions' failure to consider a holistic prioritization of projects is not due to a lack of effort or commitment by management but to the continuing assumption that research centers should choose between academic rigor and economic profitability, with no overlap.

Although both visions (academic and practitioner) may seem to contradict each other and pull in opposite directions, this case can be solved by aligning both sides to the same integrated vision, one that would include the best of both: the academic rigor and the socioeconomic relevance (Fig. 3.1).

The first action to solve the misalignment among different levels of the organization with the eternal dilemma between academic and economic metrics [1, 5] is to design a holistic group of a few key performance indicators (KPIs) to measure the ongoing progress of your organization, help you in the decision-making process and align the goals of the institution's members.

In other words, the aim is to apply performance metrics throughout the decision-making process, not only translating your institution's mission into results but also considering the academic and socioeconomic impact, thus ensuring research contributions and financial sustainability. Meanwhile, the impact of the potential results (academic, economic, and social) and the risk of implementation should be borne in mind.

Many of these principles (such as holistic overview of performance, and recurrent industry funding to increase the number of publications and entrepreneurial outputs) have been supported by several academics [6–8] and are applied by directors in different programs such as the National Aeronautics and Space Administration (NASA), the multinational conglomerate corporation General Electric, the MIT Deshpande Center for Technology Innovation, and the McKinsey Global Institute (MGI) [9].

For example, the Deshpande Center prioritizes projects in the research stage using the formula of impact divided by risk.[1]

$$Priority\ of\ project = \frac{Impact\ (i.e.,\ academic \times economic \times social)}{Risk\ (i.e.,\ feasibility \times viability)}$$

This notion of "impact" includes three proxy variables: academic, economic, and social relevance. Each proxy is measured by different metrics such as scientific advance (e.g., publications), economic income (e.g., financial resources), and improvement of society (e.g., jobs created).

The concept of "risk" includes the feasibility and viability of getting the analyzed project to an innovation stage from where it can be moved to another agent of the innovation ecosystem, such as an investor or a client. This may include the time cycle (e.g., the years needed to get to market), the cost of the research project (e.g., financial resources), and the quality of the research team (e.g., years of experience). As another example, the McKinsey Global Institute suggests a different formula with similar concepts. In this case, the impact includes two proxy variables: product contribution (e.g., the number of new products, the number of new patents, and academic advancement) and product maturity (e.g., level of development, time to impact on society, market traction) divided by the risk, which in this case is the cost of the project.[2]

$$\text{Priority of projet} = \frac{\text{Impact (i.e., product contribution} \times \text{product maturity)}}{\text{Risk (i.e., cost of research and development)}}$$

Note that the proxies—the translation of the variables to measure—should take into account the characteristics of the research center: its age (length of experience) and orientation (whether its focus is research or innovation), which relates to the center's vision and mission [10]. Let us look at a few examples.

While the proxies may include the potential number of papers published in top journals, they may also include the potential number of people attending academic conferences if the research center is young or the potential increase in citations of the center's papers if the research center is mature.

While the proxies may include the number of professional publications (such as white papers or books), they may also include the potential number of new consulting projects if the innovation center is young or new spin-offs if the innovation center is mature.

However, the proxies should always encompass a holistic perspective of the impact (i.e., academic, economic, and social) and the risk. Then the proxies should be adapted to those variables, depending on the center's specific type, vision, and mission.

Finally, the selected metrics should be made known and incentivized somehow in the whole organization to ensure the alignment of the institution to those values [11, 12]. For example, the University of Michigan shares with all the faculty a "best practices and tool kit" for the creation, management, and closing of research centers. The document sets out the university's requirements (such as for opening a center, keeping it open, and closing it) and includes suggestions of key performance metrics to track research centers, and so on.

Table 3.2 Template of a research scorecard

	IMPACT Academic (#publications)	IMPACT Economic ($ profits)	IMPACT Social (#attendees)	RISK Viability (%)	ACTION
Research - projects funded with public funds					
Apply to the public funding of the institution X1 for Y1	◕	○	◑	○	Decline
Apply to institution X2 to get local public funding for Y2	◔	○	◑	◑	Decline
Start negotiations with the institution X3 for Y3	●	◕	◑	◑	Test viability
Research - projects funded with private funds					
Publish white paper with the institution X4 about Z1	●	◔	◔	●	Start
Publish book with the institution X5 about Z2	●	●	◑	◕	Test viability
Release prototype of technology A1 to the institution X6	●	◔	○	◑	Test viability
Initiatives - new releases					
Create a open innovation competition with institution X7	○	◔	◑	◑	Test viability
Create a conference with institution X8	◕	◑	●	◕	Start
Propose a consulting project about Z3 to institution X6	○	◑	○	◑	Test viability
Initiatives - new geographies					
Expand initiative Z1 to the geography G1	◔	◑	◕	◕	Start
Expand initiative Z4 to the geography G2	●	◑	◔	●	Start

Source Prepared by the author[4]

In the end, this tool is a way of translating the mission and vision of the institution, taking into account the academic, economic, and social impact, in order to ensure research contributions and financial sustainability within society.[3]

Now let us imagine that you are preparing your research center's business plan for the next 5 years. How can you apply this principle? You will have a list of new research project activities (rows in a table) that each member of your center has transmitted to you (e.g., new opportunities, new changes, and new markets). In the columns you will have each key performance indicator. The final column will give you the recommended order of priority for the initiatives (See Table 3.2).

In the table, the blacker the circles are, the higher the value they contain. The variables are represented in columns (academic, economic and social impact, and risk) with the proxy in parentheses (i.e., potential number of new publications, new profits, new attendees to conferences or dissemination initiatives, viability/feasibility of the project). In the last column is the final decision: to decline, to test its viability, or to start it. The tool is called a research scorecard, which provides you with an easy and visual method to choose from among new initiatives and to keep track of your current initiatives.

Nevertheless, is there an easy method to calculate the viability of the project, to identify whether my researchers are already working on similar topics or to select which researchers can execute the project?

3.3 Map the Focus of Study of Each Researcher and of Your Center With the Research Map

Do you know what your research teams are investing time in? Do your researchers know what the center's other researchers are doing? Are you sure that there is no duplication among your center's research projects?

In the in-depth analysis of the research centers sampled, a large number of centers answered "no" to at least one of these three questions, especially to the second. This is a symptom of low levels of knowledge sharing within research centers (e.g., between units, between teams, and between researchers).

Managing directors at research centers usually have tight limits on resources and time (See Table 1.1) and are supported by an academic director (See Fig. 6.3). As a consequence, some of them do not invest enough time to develop an understanding of what their researchers are investing time in or they do not ensure that this work is done by the academic director. Therefore, in some cases, centers were creating duplication among research teams that were investing time in very similar topics.

Managing and academic directors at centers are able to solve this problem through mapping and connecting the focus of researchers in a research map, which illustrates on a single page the interests of each researcher and of the center.

Additionally, this map provides an illustration to identify synergies between research projects, [13] reduce the cost of duplication, improve the assignment of project needs to researcher interests, recognize collaborations with nonacademic units, improve the research strategy at the center level, and increase networking opportunities among researchers within the institution [14].

Table 3.3 Research map of Harvard Business School's entrepreneurial management unit

Interest vs Professor	Professor 1	Professor 2	Professor 3	Professor 4	Professor 5	Professor 6	Professor 7	Professor 8	Professor 9	Professor 10	Professor 11	Professor 12	Professor 13	TOTAL
alliances / joint ventures	1													1
behavioral finance						1								1
corporate entrepreneurship							1							1
corporate governance / finance / financial analysis			1			1		1	1					4
electronic markets		1												1
entrepreneurial finance						1	1				1			3
entrepreneurial management			1											1
entrepreneurship	1	1		1			1	1	1		1	1		8
foreign direct investment			1											1
government and business / economic development										1				1
innovation	1					1	1	1		1				5
international business / globalization			1			1								2
international finance			1											1
managing growth				1										1
marketing and sales force		1												1
negotiation									1					1
network organizations / networks	1		1											2
patents	1													1
patents / intellectual property							1							1
pricing			1											1
private equity	1					1						1		3
risk management									1					1
taxation		1												1
technology management / change / strategy		1		1			1							3
valuation									1					1
venture capital	1					1	1					1		4
venture creation / development		1												1

Source Prepared by the author based on an analysis of publicly available data about the school [15]. *Note* The table was simplified by excluding assistant professors, visiting professors, lecturers, postdoctoral fellows and Ph.D. students, and including only the first areas of interest of each professor. Each name has been replaced by the word "professor" to keep the anonymity

The first table depicts the current research map of the entrepreneurship unit's faculty at Harvard Business School in the United States (See Table 3.3).

The second example shows the research map of the machine intelligence research group at the University of Cambridge's Department of Engineering (See Table 3.4).

Similar methods and mechanisms to produce a comprehensive map of knowledge sources have been used by companies such as the pharmaceutical company Hoffmann-La Roche—most commonly known as Roche—and the World Bank [17].

Table 3.4 Research map of the machine intelligence research group at the University of Cambridge's Department of Engineering

Interest vs Professor/Researcher	Professor 1	Professor 2	Professor 3	Professor 4	Professor 5	Professor 6	Researcher 1	Senior research associate	Researcher 2	TOTAL
computer vision	1									1
robotics	1									1
speech and language processing		1	1		1	1				4
statistical machine translation		1				1				2
speech synthesis		1	1					1		3
machine learning		1	1		1	1				4
automatic speech recognition			1		1	1		1		4
medical ultrasound imaging				1						1
audio indexing					1					1
language modeling					1					1
spoken dialogue systems						1				1
education games							1			1
shape from photometric stereo									1	1
polarization and defocusing									1	1

Source Prepared by the author based on an analysis of public sources [16]. *Note* The table was simplified by excluding those who do not specify their area of focus and by excluding research students, research assistants, readers, and lecturers. Each name has been replaced by the word "professor" or "researcher" to keep the anonymity

In summary, this tool serves to improve the efficiency, effectiveness and interconnectivity of your center's research projects.

The final step is to make the research map accessible to the institution's senior executives of and the leaders of the research team. These knowledge-sharing methods are currently applied by organizations such as the technology companies Hewlett-Packard and Google and the public institution Health Canada, improving knowledge sharing across units, facilitating knowledge sharing through informal networking, and establishing common language and frameworks for knowledge management [17].

Finally, the process will be enhanced if the synergies among different researchers can be incentivized in some way—for instance, by including metrics in the annual evaluations to quantify the level of collaboration with other researchers through coaching, shared publications, or involvement in regular events to share best practices.

3.4 Use Professional Recruitment Processes for Academic and Executive Directors

Is your leadership team uncomfortable with managing an institution or does the team lack nonacademic experience? Is your executive team failing to understand the preferences and mindset of the institution's academics? Is it difficult for your executive team to handle academic environments? Has the academic rigor of your publications declined recently?

Three of the main unsatisfying tasks for a research center's managing director are motivating faculty members to take opportunities to interact with industry, mediating between industry and faculty when projects do not go as planned, and mediating and trying to exchange ideas that could lead to sponsored research projects. How should potential directors be identified and recruited?

In some cases, the main criterion for selecting research center directors was the time they had spent working in the institution. Although they had greater knowledge of their institution, sometimes the newly appointed directors lacked the professional experience to develop the center's economic sustainability. In other cases, the newly appointed director lacked the theoretical experience to understand the academic perspective. In both cases, this represented a bottleneck in the center's performance [18]. However, several institutions are turning to professional recruitment for leadership positions.

A former vice president of the Barcelona Supercomputing Center, for example, was a former executive at IBM with previous experience in knowledge-based projects: a perfect fit for the institution.

Another instance is Cornell Tech, the technology-focused campus of Cornell University in New York. This center recruited a professional investor with research experience to increase the number of spin-offs from research projects and to assess the different projects involving the faculty and other business units. This was someone who often complemented researchers' business knowledge.

In these cases, there was a professional recruitment process. If the perfect candidate was not in the organization, there was no problem in going to the market.

Other ways to solve the problem include splitting the director role into two specialized functions, as the School of Chemistry at the University of Melbourne does, ensuring two leadership roles—the executive director (or manager) and the academic director (or head).

Other mechanisms, sometimes applied by young research centers, include partnering with other institutions—either academic of practitioner—to fill a gap or compensate for a weakness. This practice is especially common when applying for public funds—grants that are attractive to obtain but where it is difficult to have the expertise in-house to do so.

3.5 Attract and Recruit an International Advisory Board

Do you lack either the academic or business expertise for your strategy? Do you find it difficult to assess the potential economic impact that a project may have on the industry?

Managing directors at research centers choose what they considered the three main benefits of having an international advisory board for carrying out their roles (See Fig. 3.2). The best-ranked were the design of research and sustainability road maps, in addition to an improved connection with the market through positioning and partnership referrals.

These committees have not only provided a pool of new ideas for better connections with market needs but also a channel to identify and give visibility to how those initiatives affect the external and internal ecosystem, while creating evangelizers for the institution.

Examples are the Pittsburgh Technology Council, the Petrópolis Technopole in Rio de Janeiro state [20] and the board of the Porto Digital science park in Recife, Brazil. These committees explicitly represent key local innovation actors playing a kind of a political role for enhancing local innovation capacity [21].

Similarly, the Knowledge Circle of Amsterdam meets regularly to formulate and propose ideas for enhancing knowledge-based development. "After-hours clubs in New York City can also be considered as

Fig. 3.2 Main benefits of an international advisory board. *Source* Adapted by the author from several sources such as Sander, E. Chapter 5—Industrial collaboration and innovation in *ERC best practices manual* (National Science Foundation Engineering Research Centers Program, 2013) [19]

a consensus space, providing venues for artists, fashion designers, and other creative individuals to develop new projects across arts and fashion" research institutions [21, 22].

This tool is not only useful for creating with actors from outside your institution but also with stakeholders from inside your institution, so contributing to the alignment of decision-making processes.

Endnotes

1. This formula is a simplification for explanatory purposes. Adapted by the author from the interviews conducted and complemented with several public sources such as Mission and History. *MIT Deshpande Center for Technological Innovation*. Available at: http://deshpande.mit.edu/about. (Accessed: 1st March 2017) [23].
2. Simplified formula. Adapted by the author from the interviews conducted and complemented with several public sources such as Hannon, E., Smits, S. & Weig, F. Brightening the black box of R&D. McKinsey Q. 1–11 (2015) and About MGI. *McKinsey & Company*. Available at: http://www.mckinsey.com/mgi/overview/about-us. (Accessed: 29th March 2017) [24].

3. The selection of performance metrics on how to pick the best research initiatives is not the complete solution for commercializing discoveries but tackles one of the main challenges reported by managing directors at research centers. Each center needs to define the mission, vision, and strategy. Afterwards, performance metrics will help aligning the efforts and measuring the results.
4. The examples do not refer to any existing research center but are consistent with the numbers. The variables starting with X represent names of institutions, the variables starting with Y or Z represent projects and topics, the variables starting with A represent technologies, and the variables starting with G represent geographic areas.

References

1. Campbell, T. I. D. & Slaughter, S. Faculty and administrators' attitudes toward potential conflicts of interest, commitment, and equity in university–industry relationships. *The Journal of Higher Education* **70**, 309–352 (1999).
2. Lee, Y. S. University–industry collaboration on technology transfer: Views from the ivory tower. *Policy Studies Journal* **26**, 69–84 (1998).
3. Lee, Y. S. The sustainability of university–industry research collaboration. *Journal of Technology Transfer* **25**, 111–133 (2000).
4. Nilsson, A., Rickne, A. & Bengtsson, L. Transfer of of academic research—Uncovering the grey zone author. *The Journal of Technology Transfer* **35**, 617–636 (2010).
5. Lee, Y. S. 'Technology transfer' and the research university: a search for the boundaries of university–industry collaboration. *Research Policy* **25**, 843–863 (1996).
6. Welsh, R., Glenna, L., Lacy, W. & Biscotti, D. Close enough but not too far: assessing the effects of university–industry research relationships and the rise of academic capitalism. *Research Policy* **37**, 1854–1864 (2008).
7. Bozeman, B., Rimes, H. & Youtie, J. The evolving state-of-the-art in technology transfer research: revisiting the contingent effectiveness model. *Research Policy* **44**, 34–49 (2015).
8. Gulbrandsen, M. & Smeby, J. C. Industry funding and university professors' research performance. *Research Policy* **34**, 932–950 (2005).
9. National Aeronautics and Space Administration. *NASA Strategic Plan*. (Washington DC, 2014). https://www.nasa.gov/sites/default/files/files/FY2014_NASA_SP_508c.pdf.

10. Perkmann, M. *et al.* Academic engagement and commercialisation: a review of the literature on university–industry relations. *Research Policy* **42**, 423–442 (2013).
11. Debackere, K. & Veugelers, R. The role of academic technology transfer organizations in improving industry science links. *Research Policy* **34**, 321–342 (2005).
12. Besley, T. & Ghatak, M. Competition and incentives with motivated agents. *American Economic Review* **95**, 616–636 (2005).
13. Boardman, P. C. & Corley, E. A. University research centers and the composition of research collaborations. *Research Policy* **37**, 900–913 (2008).
14. van Rijnsoever, F. J., Hessels, L. K. & Vandeberg, R. L. J. A resource-based view on the interactions of university researchers. *Research Policy* **37**, 1255–1266 (2008).
15. Harvard Business School. *Faculty—Entrepreneurial Management—Faculty & Research* http://www.hbs.edu/faculty/units/em/Pages/faculty.aspx (2017).
16. University of Cambridge. *Machine Intelligence—Group Directory |Department of Engineering* http://www.eng.cam.ac.uk/people/research-group/214?field_user_surname_value_1=&field_user_list_category_tid=All (2017).
17. Shearer, K. & Bouthillier, F. Understanding knowledge management and information management: the need for an empirical perspective. *Information Research* **8**(1), 141 (2002).
18. Prats, J., Siota, J. & Gironza, A. *2033: compitiendo en innovación* (PriceWaterhouseCoopers; IESE Business School, 2016).
19. Sander, E. *ERC Best Practices Manual Chapter 5 Industrial Collaboration and Innovation* (National Science Foundation, 2013).
20. Mello, J. M. C. de & Rocha, F. C. A. Networking for regional innovation and economic growth: the Brazilian Petropolis technopole. *International Journal of Technology Management* **27**, 488 (2004).
21. Ranga, M. & Etzkowitz, H. Triple Helix systems: an analytical framework for innovation policy and practice in the Knowledge Society. *Industry and Higher Education* **27**, 237–262 (2013).
22. Currid, E. How Art and Culture Happen in New York. *Journal of the American Planning Association* **73**, 454–467 (2007).
23. MIT Deshpande Center for Technological Innovation. *Mission and History* http://deshpande.mit.edu/about (2017).
24. McKinsey & Company. *About MGI* http://www.mckinsey.com/mgi/overview/about-us (2017).

4

Stage 2: Transformation—Translating Discoveries into Impact for the Market Through Design Thinking

Abstract This Chapter detects the four causes behind the failure to translate discoveries into inventions: ignorance regarding market need; researchers' lack of business knowledge and engagement with the industry; a scarcity of academic or executive profiles within a research team; and uncoachable researchers. Five hands-on mechanisms being applied to tackle those difficulties at prominent research centers are then presented: translating and mapping consumer needs through design thinking; following lean research principles by maximizing learning speed and minimizing testing costs; complementing the current services of the technology transfer office; creating diversified teams of academics (with Ph.Ds.) and executives (with MBAs); and measuring—in the recruitment, evaluation, and incentive scheme of academics—the ability to be mentored.

Keywords Design thinking · Lean research · Paper prototyping · Industry engagement · Investor-in-residence · Technology transfer office Market map · Linked innovation · Invention · Knowledge translation · Customer journey map · Research team diversity

It is a chilly day in Stanford, California. In classroom Lathrop 190 of the Hasso Plattner Institute of Design ("d.school"), of the prestigious Stanford University, 18 Ph.D. students are attending a recently created course for academics about how they might understand research as design.

In this course, Ph.D. students are grouped into diverse teams from different schools and with different types of expertise to solve each other's problems using specific mechanisms such as paper prototyping, customer journey mapping, and business model canvassing.

While sometimes researchers need a few months to receive feedback about their current work, these tools and mechanisms have helped these future academics to reduce the long time needed to receive feedback from other academics [1].

This is just one example of how some universities are starting to implement methodologies to improve the transformation of knowledge through faster feedback loops.

After an academic has done the research process (stage 1), obtaining certain results, the next step in the innovation funnel is the transformation phase, a stage for converting knowledge assets into tangible assets (an article in an academic journal, a prototype of a new product, a design of a new possible service, etc.). What are the bottlenecks during this process and how are they to be solved?

4.1 From Causes of Broken Innovation to Best Practices for Linked Innovation

Are you getting products no one wants to buy? Are you getting outdated products? These were the two possible symptoms identified in the case of broken innovation in market understanding during the transformation stage.

In the first case, you are probably just assuming what the market needs, without validating what the market actually wants. In the second case, you are probably just following exactly what the market says that it currently needs, without taking into account what the market will desire in the future.

Among the analyzed cases, four causes were identified as the trigger of these symptoms of broken innovation during transformation (stage 2). These four aspects are as follows: unidentified market needs, an unknown business arena, a lack of academic or executive profiles in teams, and uncoachable researchers.

First, unknown market needs. After the research stage, researchers will come up with results for a new concept, idea, or technology. The next step is the transformation from those intangible assets into tangible ones, such as a new product or service. However, what is the best product among the different choices?

In this stage, I assume that you have already applied the practices of linked innovation for the research stage, so you have adopted a holistic selection of the performance metrics (See Sect. 3.2). However, although you may have the right metrics, how are you going to reach the right decision if you do not have the right information to decide? An ineffective evaluation of the product-market fit may create this misalignment, with the assumptions of the decision process being based on mistaken expectations.

Second, an unknown business arena. While academics are specialists with narrow research questions and specific areas of expertise—sometimes the worldwide experts in their fields—they occasionally lack knowledge in other areas.

For example, a researcher who for the previous 20 years has analyzed new techniques to identify a less intrusive method to identify meningitis in newborns maybe does not know the best way to create a patent for his or her last prototype, the new legal requirements for creating and validating a prototype, the product requirements for raising public European funds for the next stage of research or how to create a profitable business model from the research. In summary, there is a lack of technical expertise or specific knowledge in areas that are outside the researcher's focus—expertise that is sometimes crucial in the transformation stage.

Third, diversity. Depending on the orientation of your center, either research or innovation, there are two types of problems with a lack of diversity. In the first case, it is common to create teams of only academics (e.g., researchers). In the second case, it is common to bring to the table a team of only nonacademics (e.g., consultants). Nevertheless, these two variations of teams usually trigger misalignments in relation to academic needs or market priorities.

For instance, although methodologies are emerging to close the rigor-relevance gap, to be academically rigorous in a project is sometimes difficult without academics on the team because performance metrics perhaps give greater weight to academic criteria, while it is sometimes difficult to be market relevant in a project with no practitioners on the team because performance metrics may grant greater weight to economic criteria [2].

Fourth, uncoachable researchers. One of the reasons why researchers follow an academic path is freedom—freedom to be involved in the projects they like, in the countries they want, at the speed they prefer. Still, one of the main concerns of research center managing directors is hiring researchers who are receptive to recommendations and suggestions—in other words, employees who are coachable. Unaligned groups of academics grouped together for one vision and mission are not only difficult to manage but also difficult to move to a process of linked innovation.

These causes were more or less common depending on the center type—its age and orientation. In the following sections, we will see the four best practices that solve each of the identified problems (See Table 4.1).

4.2 Gather Consumer Needs Through Design Thinking in the Market Map

Do you have products that no one wants to buy? Do you lack market traction when you reach the commercialization stage of your discoveries? Do your competitors identify market opportunities before you do?

According to several studies, the identification of technology customers—those who will use your inventions—is the main difficulty in commercialization [3, 4]. So is there any way to ensure that potential customers will buy your discoveries?

Table 4.1 Stage 2—transformation: translating discoveries into impact (overview)

Orientation and age of the center					
Research young	Research mature	Innovation young	Innovation mature	Causes of broken innovation	Best practices for linked innovation
x	x	x		(a) Information: unknown market needs	(a1) Use design thinking and the market map: translate and map consumer needs
					(a2) Follow lean research principles: maximize your learning speed and minimize your testing cost
x	x	x		(b) Context: unknown business arena or little engagement with industry	(b1) Complement the existing services of your knowledge and technology transfer office
x	x			(c) Diversity: lack of academic or executive profiles	(c1) Create diverse teams of executives (with MBAs) and academics (with Ph.Ds.)
	x		x	(d) Mentoring: uncoachable researchers	(d1) Include the indicator "coachable" in the recruitment, evaluation and incentive scheme for academics

Source Prepared by the author

According to the Organization for Economic Cooperation and Development (OECD), new tools and approaches such as design thinking are being applied "to manage uncertainty, and to respond to changing user demands for personalized digital services and convenient automatised processes that rival the efficiency and effectiveness of industry" [5, 6]. In this case, the combined use of design thinking and the market map may help in this endeavor.

The first method, design thinking, is a tool for innovation of both products and services. Two years ago, I explained—in a panel discussion at Harvard Business School before MBA students—that It uses a consumer-centered approach that puts the observation and discovery of highly nuanced—even tacit—consumer needs at the forefront of the innovation process. Beyond consumer needs, it considers feasibility (technological system constraints) and viability (business context) [7].

In this consumer-centered approach, a researcher or research team is able to understand and translate market needs into actionable insights and thus ultimately make informed decisions during the transformation stage.

This process allows research teams to defend effectively their own thoughts about an addressable market or test their hypothesis about a potential customer through rapid iteration and prototyping.

This human-centered, prototype-driven innovation practice integrates business and technical factors and seeks to engage multidisciplinary research teams, enabling "creative collaboration across spatial, temporal and cultural boundaries" [8]. The power of embracing a prototyping mind-set during the transformation stage allows researchers to make rapid iterations across research and product development, and subsequently adapt and learn from what does not work.

In fact, the value of this approach has been advocated by different types of organizations [9] such as IDEO in California and the United Kingdom's Design Council [10, 11]. Additionally, more institutions are claiming to apply this concept in their research processes [12].

Although some researchers are still reluctant about design thinking, others are leveraging this incredible tool. For example, as mentioned in the introduction to this chapter, Stanford University's Institute of Design has already introduced a course about how Ph.D. students can understand research as design. In this course, Ph.D. students are grouped into diverse teams from different schools and with different levels of expertise and they try to solve each other's problems using specific mechanisms such as paper prototyping, customer journey mapping, and business model canvassing [1].

For instance, during paper prototyping, in a classroom with Ph.D. students of different profiles in terms of research topics and nationalities, they try to solve the research problems of others. At first glance, skeptics think that students of different profiles cannot support each other: how can a Ph.D. student of physics studying the behavior of black holes help a Ph.D. student of engineering studying the behavior of Intel processors? However, at the end of the day, both share many common problems, such as which method is the best for addressing my research question, how can resources be found to fund my research, and how well explained is the article for a person outside the field? This methodology not only reduces the speed to develop an academic paper but also helps students gain an outsider's perspective, learn how to explain their research, and so on.

Design thinking helps not only research centers in universities and public research institutions improve the transformation stage but also research centers in industry. In a recent study, which I coauthored with the consultancy Oliver Wyman, we describe the case of the banking sector: companies such as the Industrial and Commercial Bank of China's Research Center, Auckland Savings Bank's Innovation Lab in New Zealand, J.P. Morgan Chase & Co.'s Technology Hub in the United States, BNP Paribas' Innovation Center in France, Wells Fargo's Research Center in the United States, and the Royal Bank of Canada's Innovation Center are implementing design thinking in their research centers to improve the user experience and to find new business models to sustain the growth of their business [13].

After the needs are gathered, the second tool is the market map, a method for mapping market needs, identifying patterns of necessities, and matching those patterns with research interests, through the research map (See Sect. 3.3). To create the market map, first start with the hypothesis of which type of company—geography, sector, and so on—you think is most likely to pay for your current discoveries or knowledge assets or those you have the potential to create. Then identify which companies follow those patterns and contact those businesses to validate the model.

There are many ways to gather and validate the needs of the market: search on databases such as MarketLine, which explains the current state, problems, and needs of different markets by sector and by geography; attend conferences explaining the current arena of a sector; read the strategic plans of a sector's corporations, and so on.

Another mechanism that research centers are applying is improving their technology transfer offices by having employees in those offices spend time with potential customers and leads of the center's research projects. For instance, understanding their worries, their challenges, their needs, and so on. These offices have been launched successfully in research centers such as the Barcelona Supercomputing Center—which has one of the best-performing supercomputers in Europe—and the Center for Genomic Regulation, a center with more than 400 researchers of 43 nationalities.

In these cases, their technology transfer offices were made up of specialized sales employees with the double background of an MBA and a Ph.D., profiles that aim to identify the needs and interests of both the market and the researchers, identifying, aligning, and connecting with opportunities outside the organization.

Let us look at a complete example.

The market map template includes the names of the companies of each selected sector in the columns and their topic of interest in the rows (See Table 4.2).

After identifying the interest of those companies through design thinking—by means of phone calls, focus groups, interviews, data analysis, and so on—you are able to identify which needs are those most desired in the industry in question. In this example, more companies are interested in topic 3 than in any other topic.

Next, we combine this map with the research map, which shows the interests of all the researchers, and so identify the intersections in the linked innovation map (See Table 4.3). This final map provides a lot of insights and shows some of your bottlenecks.

Table 4.2 Market map template.

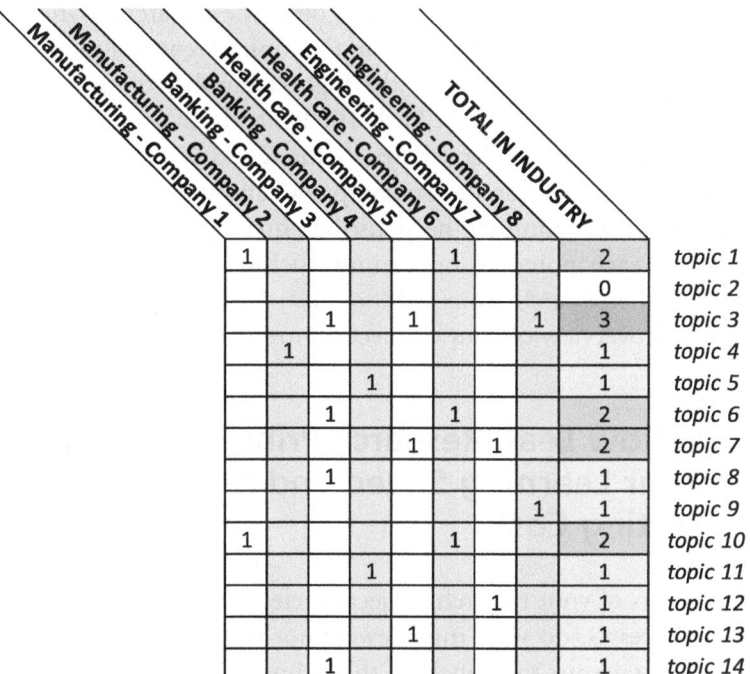

Source Prepared by the author. *Note* The company names and the topics of interest were removed for confidentiality reasons

Table 4.3 Linked innovation map: matching of research and market opportunities

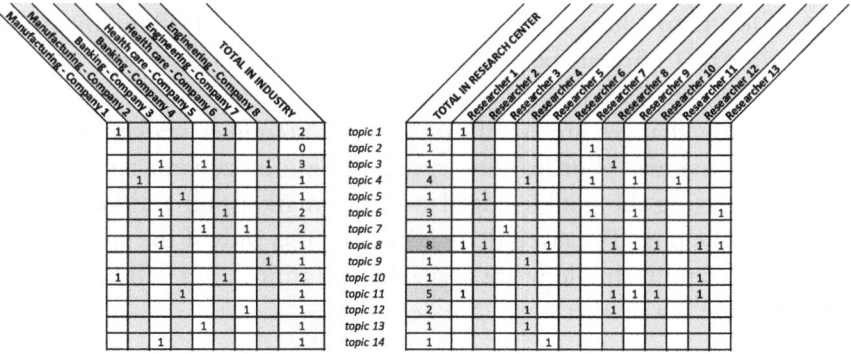

Source Prepared by the author. *Note* This map can be used to identify not only topics of interest but also the type of services (e.g., ad hoc consulting, licensing, and training)

For instance, although you know that topic 3 is the most desired among the selected companies, there is only one researcher—researcher 8—who has interest and experience in the topic. Second, although topic 8 is the one being developed by the highest number of your researchers—eight researchers: 1, 2, 5, 8, 9, 10, 12, and 13—there is only one company among those selected that is interested in the topic. Finally, topic 6 has an intersection of two companies and three researchers.

Although these numbers need further qualification with regard to the value of each potential opportunity before these values should be included in the research scorecard (See Sect. 3.2), the map provides you with a quick overview of your connection to the market.

4.3 Follow Lean Research Principles—Maximize Your Learning Speed and Minimize Your Testing Cost

Are the results of your research projects irrelevant to the market? During research processes, do your interviewees not want to repeat the process because you take up too much of their time? Do you exceed the estimated budget of your projects or increase the analysis sample to a size that does not change the conclusions?

I was shocked the first time I saw the R & D model for discovering and developing a new molecular entity. It had an estimated success rate of less than 4%, a capitalized cost per launch of $1.78 billion and a cycle time of 13.5 years before launch. Such a model keeps away inexperienced entrants (See Fig. 4.1) [14].

Since the changes are significant and the resources scarce, the need arises to maximize your learning speed and minimize your testing cost. Interconnecting the double-edged sword of market pull and knowledge push may help us. Perceived demand will be met only if the appropriate knowledge and technology are available, and innovation will be realized only if there is a market for it (See Sect. 2.4).

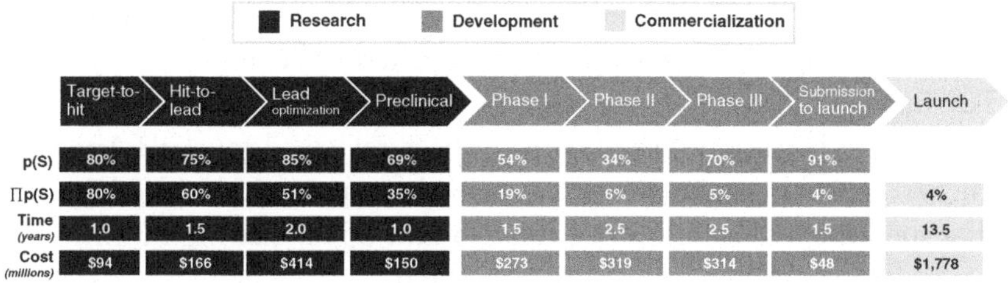

4 Stage 2: Transformation—Translating Discoveries into Impact ...

	Target-to-hit	Hit-to-lead	Lead optimization	Preclinical	Phase I	Phase II	Phase III	Submission to launch	Launch
p(S)	80%	75%	85%	69%	54%	34%	70%	91%	
$\prod p(S)$	80%	60%	51%	35%	19%	6%	5%	4%	4%
Time (years)	1.0	1.5	2.0	1.0	1.5	2.5	2.5	1.5	13.5
Cost (millions)	$94	$166	$414	$150	$273	$319	$314	$48	$1,778

Research — Development — Commercialization

Fig. 4.1 Example of an innovation funnel from discovery to launch of a new molecular entity. *Source* Prepared by the author using data from Paul et al. [14]. *Note* p(S) means probability of successful transition to the next step, $\prod p(S)$ means the productive—accumulative product—of the probability, time means the years needed per step, cost means the returns required by shareholders to use their money during the process

Research projects usually consist of multiple hypotheses. First, there is a vaguely defined hypothesis regarding the research question you are working on, which can take years to test and could become a dissertation. Second, there is a working hypothesis that you can test in a few weeks. Generally, short-term hypotheses are more concrete, while long-term hypotheses are vaguer.

These hypotheses matter because, in systems research, great work is frequently achieved via quick iteration, repeating the formulation and testing of smaller hypotheses to achieve a bigger goal. Failing quickly allows you to hone your understanding of a problem and continually evolve your set of facts and beliefs about it [15]. You should expect your hypotheses to change over time as you courageously pursue and refine them at multiple levels.

Otherwise, you could invest an enormous amount of time and money in a project that might not go anywhere. Research centers do not have infinite resources. Therefore, it is important to maximize the learning speed, while minimizing your testing cost. So, if your hypotheses are wrong, will you know this as soon as possible? If you are right, will the answer's payoff be worthwhile? Finally, are you testing your hypotheses as efficiently as possible?

Short feedback loops in research will help you as indicators of efficiency [16]. Often this feedback may prompt you to pivot your prototype but this will make the commercialization stage easier and may increase the number of previously unpublished discoveries.

For instance, how many articles do you have in your computer that ended up not being published in a top peer-reviewed journal? An analysis of 570 journals shows an average rejection rate of 62% (See Fig. 4.2), with a special concentration between 55 and 85%, bearing in mind that the average may vary depending on the field.

The good news is that the analyzed data show no clear correlation between the impact factors and rejection rates.

So if you have to publish one paper every 4 years, you will need to submit between one and two papers on average per year, depending on the acceptance rate, bearing in mind that the median time from submission to acceptance is between 100 and 150 days, depending on the impact factor of the journal [18].

Imagine, for example, that you have made a discovery and want to publish your insights. Initially you try to publish an article, after a few informal peer reviews, in a science journal. After one peer review, the

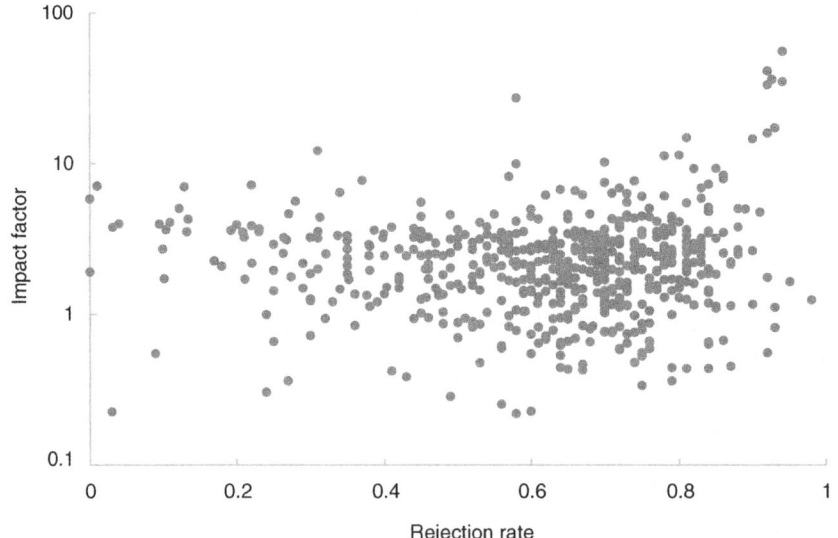

Fig. 4.2 Comparison of impact factor (in logarithmic scale) with their rejection rates ($n = 570$ journals). *Source* Adapted by the author from Rocha [17]

article gets declined. You move on to other journals, receiving a similar answer but getting the feedback that, although the research question is attractive for practitioners, it is not so attractive for academics. So you decide, rather than change the whole article, to leverage the work you have done already and submit an adaptation of the existing piece to a book proposal contest, being funded to continue your research project. After a few more iterations, you release your book with a recognized publisher, and so on. Publishing articles or transforming ideas into products is not an easy endeavor. However, pivoting through lean research techniques may help you learn what works in a more cost-effective way.

According to an article on the World Economic Forum website, lean research has been championed by faculty and researchers at MIT D-Lab,[1] the Fletcher School at Tufts University, and the Feinstein International Center, also at Tufts—with lean research being defined as rigorous, respectful, relevant, and right-sized [19].

When working in Haiti following the 2010 earthquake, Kim Wilson—a lecturer at the Fletcher School—proposed an interview template of 100 questions for a study she was conducting among survivors to learn about access to financial services. People working with survivors refused to let her ask 100 questions of people who were often traumatized and newly homeless [19]. As a result, the researcher designed a leaner way to conduct the study, making it less time-consuming for those taking part.

So how could these principles be converted into actions? The following are questions for self-reflection to check whether you are actually following the four principles outlined.

Rigorous, regardless of the methodologies employed: what steps are you taking to ensure research projects are valid? How are you designing your research to ensure it is reproducible? Will the research be verified by an independent party?

For example, a pharma investigator who replicates the findings of a clinical trial in a second external laboratory and, if possible, in a second species—a strategy that ensures robust and reproducible results [20].

Respectful, toward the research subjects: how are you protecting the participants' data? How are you helping and engaging the research subjects? How do you explain all the information to the participants in a way they can understand, while ensuring they feel truly free to take part or not? What specific steps will you take to provide participants with opportunities to review and reject the results? As in the previous case of Kim Wilson.

Relevant, to research subjects, partners, and decision-makers: what secondary research have you done in order to prioritize and ensure that primary research is actually needed on the topic you are proposing? In advance of the research project, have you identified stakeholders who have given input into how they would like to receive and use any research findings? Are you able to explain the value of the proposed research study? Have you allocated time and a budget to disseminate the research? Have decision-makers agreed about using research findings in advance of the study? How will you measure the impact of your results?

The research team of the nonprofit social investment fund Root Capital, by applying lean research principles, has implemented two changes. First, when conducting research interviews, now it asks its corporate clients (interviewees) whether they would like to include any additional relevant question in the research project. In the second change, after data are collected and analyzed, the fund shares the aggregated information with the interviewees, providing its clients with valuable information [21].

Another mechanism that other research centers are applying, following the principle of short feedback loops, is to check with practitioners and with the institution's communication department whether, if the research project is to achieve the expected results, the question and the answer are worthy [22]. In other words, are the research question and the answer relevant to the market?

Right-sized, in terms of protocols and costs compared with the study's potential usefulness and impact. In light of Kim Wilson's experience in the Haiti research project, what criteria are you using to assess how large (in terms of sample and methods) and costly it is reasonable for the study to be? How are you assessing which activities are essential to the research goals and which you can eliminate? [23].

4.4 Complement the Current Services of Your Knowledge and Technology Transfer Office

Do you have researchers who lack business knowledge or commercialization experience? Does your research team invest too much time in nonacademic issues? How easy is it for your research team to network with the relevant industry to gather data, invite guest speakers, disseminate their discoveries, etc.? Do the legal requirements of intellectual property constitute a barrier for the commercialization or publication of knowledge?

Since 2005, 34 successful spin-offs and more than 500 research projects have been created via Humboldt Innovation GmbH, a spin-off and wholly owned subsidiary enterprise of Humboldt University in Berlin, Germany. The unit acts as a comprehensive interface between the university and industry. It positions itself between business and academia [24].

Knowledge and technology transfer units optimize the time invested by academics in what they do best—research and teaching—by leveraging complementary skills and resolving the problem of lack of knowledge in specific areas unrelated to the research topic—topics related to the law, recruitment, sales, communication, etc. For instance, traditional technology transfer programs in academic institutions have relied largely on a licensing model, whereby institution license founders and research teams provide access to the institution's technology for a set percentage of future revenues during a designated product's life cycle or until the subsequent dilution of share ownership during the next funding round. If I am a researcher with a new discovery in this program, what is the best deal for the institution and for me?

Other challenges or questions this unit may answer are: what are the legal requirements for testing a new vaccine that the team has developed? How do I find a cofounder for a spin-off or an additional team member who has a better perspective of what the market wants? How do I translate my recent discovery into words that can be understood not only by other experts in the field but also by nonexperts? How do I network or get access to possible distributors or potential buyers? How do I manage and negotiate the intellectual property of the recent discovery?

All these questions can be answered by using an improved knowledge and technology transfer office, which will mentor, improve, and accelerate those processes. These functions can be done internally or externally by outsourcing.

Several centers are already applying this infrastructure. For example, I was in the facilities of the Harvard Innovation Lab, which offers services to the Harvard community such as coaching through entrepreneurs in residence, investors in residence, legal partners, visiting practitioners, experts, etc. [25].

IESE Business School launched an initiative in 2016 called the Barcelona Technology Transfer Group, which connects individuals from research and development labs, companies, and educational institutions. The initiative was an attempt to identify and mix the skill sets required to prototype and commercialize products of high value to society more successfully.

In other words, the platform seeks to bring together profiles from R & D, business, etc., to accelerate the process of venture building [26]. This program also provides enough time for the team to get to know each other in a low-risk, high-reward environment, avoiding conflicts of interest between parties.

I got into the elevator up to the 51st floor where there is the SAP Next-Gen, a corporate innovation lab—in Manhattan—that is succeeding in communicating the company's new technologies across the globe, engaging the innovation community around the company through its worldwide partnerships and university alliances, connecting more than 3,100 institutions in 106 countries [27].

Likewise, I was walking through the beautiful gardens at the University of Pennsylvania's Wharton School to visit the Mack Institute for Innovation Management, which provides its faculty with nonacademic services such as the selection and engagement of speakers for panel discussions and coaching for researchers on how to communicate and network with the market [28].

Knowledge and technology transfer offices may also help in areas such as the creation of prototypes to conduct user testing for commercialization, ensuring a product-market fit prior to a formal launch. Moreover, these units may gather information about market opportunities or market intelligence by engaging with go-to-market experts, such as the pharmaceutical company Buckman Laboratories does through a customer information center, a database containing all the information available at the company about each customer and the market [29–31].

Other examples of best practice include sharing proven mail templates for contacting a senior leader to increase the readership rate or for inviting a guest speaker to give a lecture. Another example is nurturing a CRM with potential industry collaborators (e.g., as a sponsor, as a guest panelist, as a source of data in exchange for knowledge, as a source of practitioner expertise, etc.). These tools will help, especially female or junior faculty members or those with nonacademic status—three groups that usually report low levels of engagement with industry [32–34].

An example of how research centers prepare their researchers is the team of investors in residence and developers in residence that has been built up in Cornell University's Cornell Tech campus, in operation since 2012. The group helps faculty and researchers communicate with industry, in addition to developing their ideas and connecting them to the market.

4.5 Creating Diverse Teams of Executives (with MBAs) and Academics (with Ph.Ds.)

Are your teams formed of only academics or only nonacademics? Do researchers and executives understand each other in terms of language, performance metrics, timing, and mind-sets?

The personal attributes of research leaders are found to have a profound impact on the extent of academic-industry collaboration. Faculty members with senior status and administrative positions are shown to be likelier to have more resources available for R & D activities and thus find it easier to interact with larger corporate firms in science-based sectors. Additionally, prior collaborative industry experience, gender, and the research fundraising priorities of faculty members also have a noticeable impact on the transformation process in academic research centers [35].

Additionally, it is already known that team diversity increases performance. For example, gender diverse and ethnically diverse organizations are 15 and 35% more likely, respectively, to financially outperform those that are not [36]. However, one of the bottlenecks we identified in the transformation stage is the lack of diverse backgrounds among research teams.

Research centers such as the Center for Genomic Regulation in Barcelona are already achieving a mix of research team members from different specializations (e.g., people with chemistry Ph.Ds. and those with industrial Ph.Ds.) and from different geographic areas (e.g., from Russia to Austria to the United States). However, in the transformation stage, some research teams lack a vision of what the market wants or they fail to bear in mind the forthcoming commercialization step.

Some centers have started to incorporate MBA graduates in their transfer units or even in their research teams such as in the Barcelona Supercomputing Center. The other option has been to recruit hybrid profiles—people with both an MBA and a Ph.D. Such profiles understand the language, concepts, and goals of both sides—the academics and the executives.

One successful example is IBM's $90 million nanotechnology center in Zurich, which opened in 2011. The Binnig and Rohrer Nanotechnology Center is part of a 10-year strategic agreement between IBM and the Swiss Federal Institute of Technology (ETH Zurich). The goal of the partnership in nanoscience is to advance energy and information technologies [37].

Another example is the partnership between Telekom Innovation Laboratories, a collaboration between Deutsche Telekom AG and the Technical University of Berlin. During this collaboration, several examples of best practice were applied, such as the inclusion of hybrid profiles who had a natural interest in the application of work oriented toward R & D and who understood both the academic and practitioner environments [38].

4.6 Include the Indicator "Coachable" in the Recruitment, Evaluation, and Incentive Scheme of Your Researchers

Do your researchers welcome and follow your suggestions? Do they follow the research center's strategy? Are they aligned with the center's vision and mission?

The MIT Deshpande Center for Technological Innovation announced recently that it was awarding $1.15 million in grants to 15 MIT research teams already working on early-stage technologies.

These projects were selected using several criteria. In addition to being projects with the potential to make a significant impact on quality of life, they were formed by researchers who were coachable. Researchers were eager to participate in the program and to form a partnership early to increase the likelihood of commercialization. These are criteria that the center prioritizes [39].

As an analogy, what would happen in a healthy body if the heart said, "I want to be the head" or the liver said, "I want to be a hand." Although I am not a physician, I assume that the results would be dreadful.

As with the body, a research institution in which each academic goes in a different direction, with no overall strategy or guidelines, entails several risks such as duplication, cannibalism among initiatives of the same institution's team members, or the center being economically unsustainable. How are you going to mentor researchers who have no business experience but want to commercialize their ideas, if they do not want to be coached?

A predisposition to be mentored is compatible with freedom, which is also important for transferring discoveries to the market because an academic researcher's personal efforts to create collaboration opportunities for his or her laboratory are more effective than institutionalized transfer mechanisms [40].

In other words, it is important to ensure that researchers can be coached and aligned to a strategy. This characteristic can be enhanced by including performance metrics related to this topic during the recruitment process and during annual performance reports, with these metrics having an impact on the incentive scheme.

The final step, after ensuring the researchers are coachable, is to provide them with mentors who, for instance, can educate research teams on the process of commercializing technologies.

For example, the MIT Deshpande Center incorporated a venture mentoring service that provides expertise and oversight for on-site research initiatives. The center's innovation arm provides research teams with guidance during the transformation stage as they move toward commercialization. Research experts and entrepreneurs in residence work alongside faculty members and Ph.D. graduates in order to facilitate the transformation process of research initiatives using a design thinking process. Mentors spend time with the team members, helping them make progress and figure out what to do [41].

The other example is the economic development office at Johns Hopkins University, which staffs former venture capitalists. Its mission is to support the faculty as they think about, prepare, and advise on the opportunity for commercialization of Hopkins technologies.

In summary, we should take a twofold approach. On the one hand, there should be coachable researchers who will listen to the mentors' suggestions. On the other hand, coaches should be available to guide the research team, ensuring the center has an overall aligned strategy.

Endnote

1. An initiative of the Massachusetts Institute of Technology that promotes "development through discovery, design, and dissemination."

References

1. Kolawole, E. A place for design thinking in academic research. *Stanford University Institute of Design—The Whiteboard* (2015). http://whiteboard.stanford.edu/blog/2015/03/24/a-place-for-design-thinking-in-academic-research.
2. Rynes, S. L. Let's create a tipping point: what academics and practitioners can do, alone and together. *Academy of Management Journal* **50**, 1046–1054 (2007).

3. Lichtenthaler, U. External commercialization of knowledge: review and research agenda. *International Journal of Management Reviews* **7**, 231–255 (2005).
4. Perkmann, M. *et al.* Academic engagement and commercialisation: a review of the literature on university–industry relations. *Research Policy* **42**, 423–442 (2013).
5. OECD. *Science, Technology and Innovation Outlook 2016* (Brussels, Belgium: OECD, 2016).
6. Daglio, M. *Public Sector Innovation: The Journey Continues* (Brussels, Belgium: OECD, 2016).
7. Gruber, M., de Leon, N., George, G. & Thompson, P. Managing by design. *Academy of Management Journal* **58**, 1–7 (2015).
8. Plattner, H., Meinel, C. & Leifer, L. *Design Thinking: Understand-Improve-Apply* (Heidelberg, Germany: Springer, 2011). (Source: https://hpi.de/fileadmin/user_upload/fachgebiete/meinel/papers/Book_Chapters/Front_Matter_-_Design_Thinking_Understand__Improve__Apply.pdf).
9. Siota, J. & Zorzella, L. *Revenue Growth: Four Proven Strategies—Lean Principles Applied to Growth Companies and Startups* (Madrid, Spain: McGraw-Hill, 2014).
10. Martin, R. *The Design of Business: Why Design Thinking is the Next Competitive Advantage* (Boston, USA: Harvard Press, 2009).
11. Beverland, M. B., Wilner, S. J. S. & Micheli, P. Reconciling the tension between consistency and relevance: Design thinking as a mechanism for brand ambidexterity. *Journal of the Academy of Marketing Science* **43**, 589–609 (2015).
12. Glen, R., Suciu, C. & Baughn, C. The need for design thinking in business schools. *Academy of Management Learning and Education* **13**, 653–667 (2014).
13. Siota, J., Klueter, T., Staib, D., Taylor, S. & Ania, I. *Design Thinking: The New DNA of The Finacial Sector* (IESE Business School; Oliver Wyman 2017).
14. Paul, S. M. *et al.* How to improve R & D productivity: The pharmaceutical industry's grand challenge. *Nature Reviews Drug discovery* **9**, 203–214 (2010).
15. Siota, J. & Zorzella, L. Learn fast, design and never assume you're right: the three principles of fast-growth companies. *LSE Business Review* October (2015). http://whiteboard.stanford.edu/blog/2015/03/24/a-place-for-design-thinking-in-academic-research.
16. Bhamu, J. & Singh Sangwan, K. Lean manufacturing: literature review and research issues. *International Journal of Operations & Production Management* **34**, 876–940 (2014).
17. Rocha, P. Selecting for impact: new data debunks old beliefs. *Open Science and Peer Review* (2015). http://whiteboard.stanford.edu/blog/2015/03/24/a-place-for-design-thinking-in-academic-research.

18. Powell, K. Does it take too long to publish research? *Nature* **530**, 148–151 (2016).
19. McKown, L. *Why International Development Should Use Lean Research* (World Economic Forum, 2015). http://whiteboard.stanford.edu/blog/2015/03/24/a-place-for-design-thinking-in-academic-research.
20. Lapchak, P. A., Zhang, J. H., Noble-Haeusslein, L. J. & Lapchak, P. A. RIGOR guidelines: escalating STAIR and STEPS for effective translational research. *Translational Stroke Research* **4**, 279–285 (2013).
21. MIT D-Lab. *Lean Research* (Working Paper) (2015).
22. Cohen, D. J. The very separate worlds of academic and practitioner publications in human resource management: Reasons for the divide and concrete solutions for bridging the gap. *Academy of Management Journal* **50**, 1013–1019 (2007).
23. Hoffecker, E., Leith, K. & Wilson, K. *The Lean Research Framework: Principles for Human-Centered Field Research* (MIT D-Lab, 2015). http://whiteboard.stanford.edu/blog/2015/03/24/a-place-for-design-thinking-in-academic-research.
24. Opinno. Leading Global Ecosystems Report 2013. (2013). http://whiteboard.stanford.edu/blog/2015/03/24/a-place-for-design-thinking-in-academic-research.
25. The Harvard Innovation Lab. *Office Hours—Entrepreneur in Residence and Mentor* https://i-lab.harvard.edu/meet/office-hours/ (2017).
26. IESE Barcelona Technology Transfer Group (BTTG). http://www.bcntech.eu/ (2017).
27. SAP. *SAP University Alliances|Shaping the Future of Education* http://www.sap.com/training-certification/university-alliances.html (2017).
28. Wharton School of the University of Pennsylvania. Mack Institute for Innovation Management https://mackinstitute.wharton.upenn.edu/ (2017).
29. Shearer, K. & Bouthillier, F. Understanding knowledge management and information management: the need for an empirical perspective. *Information Research* **8**, (2002).
30. Rynes, S. L., Giluk, T. L. & Brown, K. G. The very separate worlds of academic and practitioner periodicals in human resource management: implications for evidence-based management. *Academy of Management Journal* **50**, 987–1008 (2007).

31. Latham, G. P. A speculative perspective on the transfer of behavioral science findings to the workplace: "The times they are A-changin'". *Academy of Management Journal* **50**, 1027–1032 (2007).
32. Klofsten, M. & Jones-Evans, D. Comparing academic entrepreneurship in Europe—the case of Sweden and Ireland. *Small Business Economics* **14**, 299–309 (2000).
33. Bozeman, B. & Gaughan, M. Impacts of grants and contracts on academic researchers' interactions with industry. *Research Policy* **36**, 694–707 (2007).
34. Ponomariov, B. & Craig Boardman, P. The effect of informal industry contacts on the time university scientists allocate to collaborative research with industry. *The Journal of Technology Transfer* **33**, 301–313 (2008).
35. Azagra-Caro, J. M. What type of faculty member interacts with what type of firm? Some reasons for the delocalisation of university–industry interaction. *Technovation* **27**, 704–715 (2007).
36. Hunt, V., Layton, D. & Prince, S. *Why Diversity Matters* (McKinsey Quarterly, 2015). http://whiteboard.stanford.edu/blog/2015/03/24/a-place-for-design-thinking-in-academic-research.
37. Edmondson, G., Valigra, L., Kenward, M., Hudson, R. L. & Belfield, H. Making industry–university partnerships work: Lessons from successful collaborations. *Business Innovation Board AISBL* 1–52 (2012).
38. Rohrbeck, R. & Arnold, H. M. Making university–industry collaboration work-a case study on the Deutsche Telekom Laboratories contrasted with findings in literature. In *ISPIM Annual Conference: Networks for Innovation* (2007). http://whiteboard.stanford.edu/blog/2015/03/24/a-place-for-design-thinking-in-academic-research.
39. MIT Deshpande Center for Technological Innovation. *Selection Criteria*. https://deshpande.mit.edu/criteria (2017).
40. Van Dierdonck, R., Debackere, K. & Engelen, B. University–industry relationships: How does the Belgian academic community feel about it? *Research Policy* **19**, 551–566 (1990).
41. MIT Deshpande Center for Technological Innovation. *Mission and History*. http://deshpande.mit.edu/about (2017).

5

Stage 3: Commercialization—Designing Collaborative Business Models for University-Industry-Government Relations

Abstract With regard to the failure to achieve industry collaborations, this chapter uncovers seven causes: an unclear business model; a center's lack of brand; a lack of experience in the research team; an unclear value proposition; a disproportionate research team size; a center's internal bureaucracy or politics; and the unacceptance of research results by external stakeholders. Later, 12 business models being applied at high-performing research centers are presented, including technology transfer through public funding, transfer pricing, marketing collaborations, freemium, licensing, spin-offs, search models, and consultancy joint ventures. Finally, 10 practical mechanisms are provided for optimizing university-industry-government models and facing the aforementioned barriers: designing a collaborative business model that fits the center's orientation and age; reviewing the processes of your communication unit, ensuring a map of roles, processes in cascade and a CRM of specialized media; and doing periodic industry lectures to translate research results into impact; and more.

Keywords Collaborative business model · Discovery commercialization Transfer pricing · Freemium · Licensing · Technology transfer ·

Research contracting · Spin-off · Search model · Research ad hoc · Consulting · Academic marketing · Linked innovation · Research team size · Brand architecture · Public funding · University-industry-government

In one of my trips to Boston, I met a Harvard University professor who told me "you should talk with Thatcher Bell in New York," an investor in residence at Cornell Tech. At a first glance, I thought the role would be an investor of tech start-ups within the university. However, the professor surprised me when he explained that this investor gives advice not only to start-ups but also to research teams, Ph.D. graduates, and spin-offs about how to connect with and understand the market, and how to raise funds for their projects.

Of the 61 interviews I did with people in leadership roles at research centers and the surrounding ecosystems, I particularly remember the above conversation. It was a refreshing way to understand how research centers are applying corporate mechanisms in research teams, such as the common phenomenon of the investor in residence or entrepreneur in residence, not just to help MBA students' start-ups but also academics' research projects.

Cornell Tech has established a thriving supportive ecosystem for the commercialization of its internal ventures and research projects. In terms of human capital, former and current entrepreneurs, experts in the field, help to develop the initial concepts of incubated projects into a minimum viable product.

In this case, three full-time "floating" developers help to provide in-depth expertise about product architecture, software engineering, and data science. Additionally, technical specialists are on hand to provide assistance throughout the commercialization process to design growth strategies, iterate business models, and assess initial contracts.

These industry mechanisms are now proliferating in university and public research centers.

When the transformation process (stage 2) finishes, a tangible asset should be obtained, converted into a product or a service. The next step in the innovation funnel is the commercialization phase, a stage to capture economic value from the tangible assets. So what are the bottlenecks during this process and how can they be overcome?

5.1 From Causes of Broken Innovation to Best Practices for Linked Innovation

Are you finding it increasingly difficult to monetize your knowledge assets? Are you having more trouble getting access to industry data and networks?

In the first case, you are probably doing only research collaborations, with industry, to nurture your research. In the second, you are probably seen as a furtive sales researcher.

Many research centers fail to capture the full value of their discoveries because of inattention to commercialization. For instance, they obtain less than 0.5% of their operating income from licensing, although often 5–10% would be possible [1].

Of the analyzed cases, eight causes were identified as the trigger of the symptoms of broken innovation in commercialization (stage 3). These eight aspects are as follows: an unclear business model, a lack of a brand, a disproportionate size of research team, incomprehensible products, internal bureaucracy, nonacceptance of generated results, and lack of public funding.

First, the definition of a clear and validated business model is necessary to make the center economically sustainable.

Second, a lack of a recognizable research center brand or a lack of experienced researchers is a problem. Brand strength has a direct impact on the research center's market position. It is usually easier for mature research centers to attract world-class talent [2], secure partnerships, and collaborate with industry-leading enterprises, leveraging their market position.

Third is unsellable or incomprehensible products. Do you remember the first time you tried to explain to an outsider what you were doing in your center or as part of your research team? Discoveries in science are usually difficult to understand [3] and, as a consequence, difficult to sell.

A fourth cause is a disproportionate size of research team. We have found that, in several geographic areas, research centers are formed by large teams of researchers while the relevant industry in that geographic area is not big enough to interact with the teams. For instance, many Spanish research centers use large teams, while 99.9% of the market consists of small and medium enterprises with problems in accessing that knowledge or with no business units that are prepared to receive the transferred knowledge or innovation from research teams.

Fifth, internal bureaucracy and politics are a problem. In mature research centers, bureaucratic and political barriers were strong bottlenecks in the commercialization stage.

For example, do you remember the last time you proposed an industry collaboration to the leadership team but received no answer or received an answer only after a long delay? Then, when you asked why, you got an answer along the lines of "you know, internal politics…."

Moreover, do you remember the last time you were involved in a process to secure industry collaboration with your center but it was delayed or even stopped because of the tons of approvals, triple checks, and other processes you needed to comply with? Furthermore, do you remember the last time you were deciding on your research center's strategy and started to think: what is really my vision and mission? How I am going to be measured? Who is going to measure my performance? Only one person or different profiles within the institution?

Sixth is nonacceptance of generated research results. In ad hoc research projects involving external institutions, we usually deliver a document, a prototype or a service. However, since KPIs from research centers and from the contractor may be different, misunderstandings may happen when an unexpected result is received or shared, when the scope of the project expands or when a document is delivered that does not answer the specific question directly.

Seventh is lack of public funding. On the one hand, there are an estimated 15 million Ph.D. graduates worldwide, one in every 500 people.[1] The majority of them come from the United States, Germany, the United Kingdom, and India [4]. The number of Ph.D. holders is increasing at a rate of between 20 and 46% per year, depending on the country, so there is a wider variety of researchers.[2]

5 Stage 3: Commercialization—Designing Collaborative Business …

Table 5.1 Stage 3—commercialization: designing collaborative business models (overview)

Orientation and age of the center				Causes of broken innovation	Best practices for linked innovation
Research young	Research mature	Innovation young	Innovation mature		
x	x	x		(a) Business modeling: not clear or undefined	(a1) Understand and design a clear collaborative business model (a2) Select the collaborative business model that fits the orientation and age of your center
x		x		(b) Branding: lack of a brand or lack of an experienced research team	(b1) Partner with complementary brands and write mediatic reports (b2) Review the processes of your communication unit: function map, communication processes in cascade and CRM of specialized media (b3) Partner with visiting researchers or reward recognized faculty
x	x		x	(c) Transfer: unsellable or incomprehensible products	(c1) Do periodic lectures to industry translating research results into qualified impact
	x			(d) Transfer: disproportionate size of the research team	(d1) Adapt team size of the teams to the market needs
	x		x	(e) Transfer: internal bureaucracy and politics	(e1) Identify the decision makers and identify their key performance indicators
x		x		(f) Transfer: non-acceptance of generated research results	(f1) Define the delivery requirements or deadlines before starting, and presell your solution
x		x		(g) Transfer: lack of public funding	(g1) Create a specific unit to apply for public funds and leverage external research incubators

Source Prepared by the author

On the other hand, public funds have a low acceptance rate. In the first 2 years of Horizon 2020, for example, only 11.8% of proposals were accepted out of the more than 76,400 eligible proposals received, with 49% of participants being newcomers [5]. In other words, depending on your geographic area, on average you should submit between 7 and 11 proposals to get one approved, to secure a grant of €1.77 million, which varies according to the type of program.[3] There is a higher number of academics on the supply side while, on the demand side, it is increasingly challenging to succeed with applications for public funding.

Eighth is a lack of identified industry collaborations to raise private funding for your research and initiatives because you do not know either who is in charge of the budget or how to attract funds from the budget.

These causes were common to a greater or lesser extent, depending on the center type according to age and orientation. In the following sections, we will look at the 10 best practices and 12 business models to solve each of the identified problems (See Table 5.1).

5.2 Design a Clear Collaborative Business Model

Do you not know how to monetize your inventions? Are your research collaborations only a matter of gathering data? Are you seen as a furtive research seller by other stakeholders? Do you find it difficult to sustain long-term collaboration with industry?

The trend across the whole OECD area is that university research is increasingly funded by private companies and that the share of basic funding[4] for universities is decreasing. For instance, the private funding rate in Norway went from 2.8 to 6.0% [6] (See Fig. 5.1).

Establishing long-term collaborations with industry, the government, or universities (e.g., data collection, product development, and implementation perspective) leads to the creation of business models that are of mutual benefit economically, hence fostering win–win collaborations.

5 Stage 3: Commercialization—Designing Collaborative Business ...

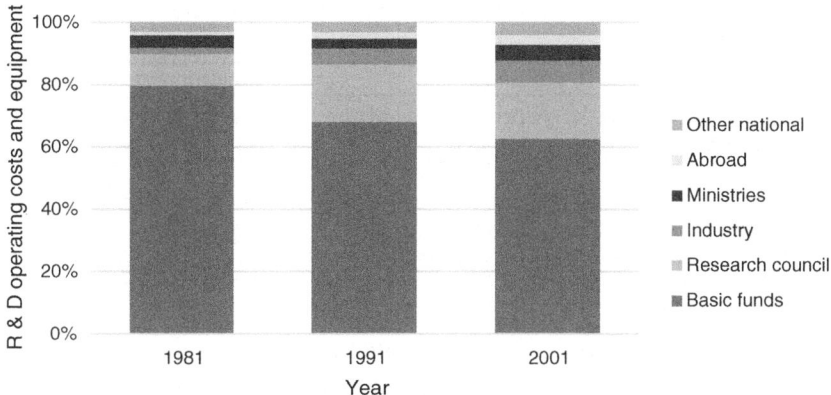

Fig. 5.1 R & D operating costs and equipment in universities by source of funding (Norway). *Source* Prepared by the author using data from Gulbrandsen and Smeby [6]

Business model design is a key decision for a new firm entrepreneur and a crucial—perhaps more difficult—task for managers charged with rethinking an old model to make their firm fit for the future [7]. This statement also applies to research center directors.

So how they can interact with each other? In the following sections, we will see in more detail how interconnected the three groups of actors are within the innovation ecosystem: corporations, universities, and governments (See Sect. 6.2).

The first step, that of designing the interactions, is to understand the benefits for each actor and to identify the business model that best fits the characteristics of your innovation ecosystem (See Sect. 5.3). In other words, it is a matter of understanding better what you and the other actor can offer—whether you are a research center at a university, in industry, or in the government (See Table 5.2).

After gaining an understanding of the potential benefits for each stakeholder, the next step is to choose the collaborative business model that best fits your characteristics.

Although more business models were identified (e.g., hybrids combining features of several models), the 12 most commonly applied at centers were short-, medium-, and long-term external contracting,

Table 5.2 Potential benefits from interactions among universities, industry, and government

Universities	Industry	Government
Link with real business	Access to state-of-the art research	Competitiveness
Source of funding		National growth
Case studies	Cost sharing	Innovation
Access to facilities	Specific expertise	Knowledge-based economy
Leveraging government funds	Training	
	Access to facilities	
Consultancy income	Joint ventures and start-ups	
Demonstrating impact		

Source Adapted by the author from Schofield [8]

internal contracting through transfer pricing, freemium product or service, research licensing, technology transfer by public funding, creation of spin-offs from the research center via external investment, the search model, the consultancy joint venture, short- and long-term marketing collaboration.

The frequency of application of each model varied with regard to several variables (e.g., recognized brand, orientation). The models with the highest estimated frequency of application were long-term external contracting (29%), short-term marketing collaboration (28%), medium-term external contacting (14%), and research licensing (7%) [9].[5]

5.2.1 Short-Term External Contracting

Every year, the Innovation Research Program of Hewlett-Packard's HP Labs research group solicits ideas from academics on selected research topics with the aim of building new research collaborations in exchange for modest grants ($50,000–$75,000).

HP receives around 500 proposals per year, selecting 10% of them on the basis of its own needs. To manage the projects' intellectual property, HP Labs uses a standardized collaborative research agreement that entitles academics to publish their results and that provides HP with the nonexclusive right to use the findings [10].

In short-term external contracting, an institution (i.e., a corporation, government agency or university) spots a problem that needs to be solved urgently using specific knowledge or a search. Then the institution identifies the research center that best fits the challenge. The center undertakes an ad hoc research project or consultancy in exchange for a fee or commission on the final result (See Fig. 5.4).

With this kind of collaboration, there is bad news and good news. The bad news is that it is difficult to attract institutions if the research center is young.

The good news is that many sectors—such as technology, banking, and retail—are increasingly aware of the need for evidence-based management. Additionally, more corporations are outsourcing or getting support from experts at universities (e.g., the Finnish technology company Nokia is in a partnership with UC Berkeley's California Center for Innovative Transportation [10]) because almost all sectors are currently reshaping their boundaries and their internal rules. These characteristics of the industries have been redesigned and disrupted by new entrants (e.g., fintech start-ups in the financial sector).

These contracting models can be implemented using several mechanisms—for instance, via an ad hoc training program for knowledge transfer, producing a document with answers to the problem set out by the company, or delivering a prototype of a new product.

In all cases, it is important to keep the research aligned with what is practical for a client organization to implement, through precisely targeted research and analysis [11]. Why? Although research projects may result in major outcomes (60%), they are not always followed by a major impact on the contractor (20%) (See Fig. 5.2).

Therefore, it is important to establish a strong relationship with the client, providing direct instructions on what information is required to develop further actionable insights that can be leveraged and applied across its product line.

These contracting projects might include gathering data and insights from x years of research and y data points, understanding or predicting the transformation of a specific subject or market that a client has asked for, or drafting a proposal for an R & D-driven innovation to change how the client thinks.

Several experiments show key aspects that differentiate successful from unsuccessful contracting projects. Research centers that carried out successful projects indicated the following as the top critical factors that

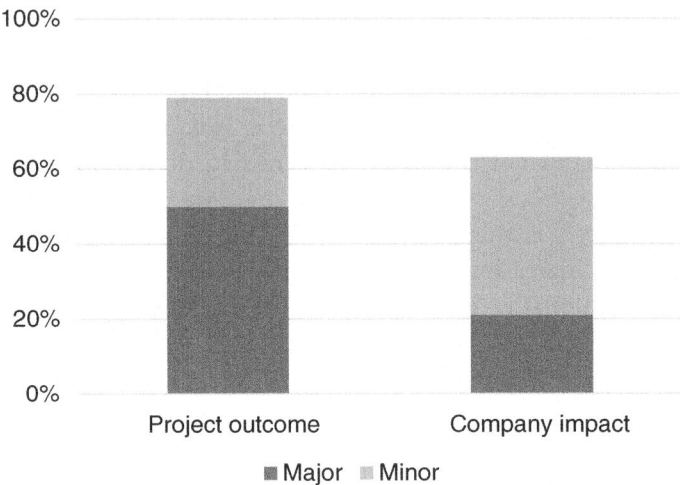

Fig. 5.2 The gap between R & D projects' outcome and company impact. *Source* Prepared by the author using data from Pertuzé et al. [12]

made them succeed: ensuring the client is really interested and involved in the project (80%), the client's capacity to assimilate new knowledge (60%), and a confident attitude toward the research group (73%).

In contrast, centers that carried out unsuccessful projects designated the following as the top critical factors that made them fail: a low level of feasibility of executing the project (53%), a high technical risk to achieve the desired results (40%), and unclear initial objectives (27%) [13].

5.2.2 Medium-Term External Contracting

The project was a success: on time and within budget, a common statement among ad hoc projects. The two boundaries are deadlines and budget.

Although similar to the short-term equivalent, medium-term external contracting differs in the planning. In this case, the company is aware much in advance of the amount of projects that the research center is going to execute in exchange for fees and resources. A collaboration may last around 2–4 years (See Fig. 5.4).

This case is useful to the institution (i.e., corporation, government agency, or university) to pursue a mix of in-house and outsourced research, helping corporations to balance their R & D portfolios, spread the risk associated with doing research in-house, and giving them access to knowledge from outside (See Table 5.2). Mathematical models in economics, finance, or social science can help and be incorporated into the internal processes of hedge funds, for example.

These collaborations are done not only by research centers at universities but also in government and corporations in sectors such as banking (e.g., the Deutsche Bank Research group in Germany), the media (e.g., the Economist Group in the United Kingdom), technology (e.g., Jigsaw—the former Google Ideas—in the United States, the Samsung Economic Research Institute in South Korea), and consulting (e.g., the A.T. Kearney Global Business Policy Council in the United States).

5.2.3 Long-Term External Contracting

In the case of long-term external contracting, the corporation can provide funding, new equipment, access to data, etc., in exchange for research capabilities. Data sharing from industry is an increasing success factor in research projects because data-driven research has grown rapidly in the past two decades, as we can see in the percentage of published articles applying data mining [4] (See Fig. 5.3).

In this case, the research center gets more involved with the corporation and may establish strong relations with executives and gain a better understanding of the company's politics. This is a kind of a long-term joint venture (See Fig. 5.4).

One example is the joint venture announced by Google and PwC in 2014. Although funds were not transferred directly in this case, as would happen with a contractor, the two companies agreed to share their core capabilities to bring further innovation to industry by leveraging PwC's business insights along with Google tools and using PwC's

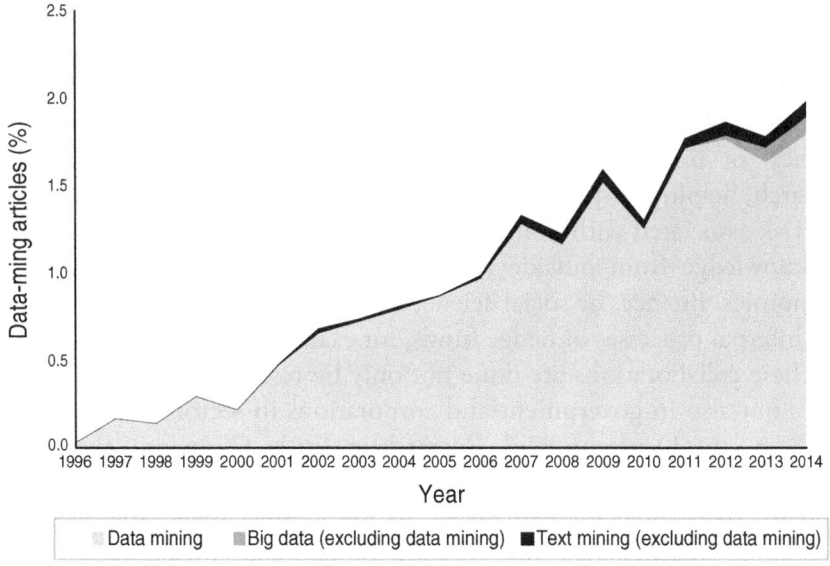

Fig. 5.3 Data mining-related scientific articles per 1,000 articles. *Source* Adapted by the author from OECD [14]

*In this case, **institution** refers to a corporation, government agency or university.*

Fig. 5.4 Short-, medium-, and long-term external contracting. *Source* Prepared by the author. *Notes* (a) The difference between short-, medium-, and long-term external contracting is that the contract fee, in the *right arrow*, and research, in the *left arrow*, are (1) ad hoc, (2) planned, or (3) sustained, respectively. (b) The most common institution in short- and medium-term contracting would be a corporation or a university. However, government agencies also contract research centers through public calls. (c) Universities and government are less common as targets in long-term contacting. Government contracting depends on election cycles

analytical acumen in conjunction with products such as the Google Cloud Platform [15].

Another example is the Grand Challenge on Clean Fossil Fuels, a collaboration between Shell and two departments at Imperial College London. It ran for 5 years, from 2007 to 2012, with Shell providing £3 million of funding [16].

5.2.4 Internal Contracting Through Transfer Pricing

Although it is more common among corporate research centers, this model is also applied in university research centers. It could be applied, for example, in another department in the same corporation or another school in the same university.

These collaborations are of great benefit to the institution because they preserve the budget internally—usually through internal vertical integration. Some institutions decide to use transfer pricing,[6] while others do not (See Fig. 5.5).

Success stories include that of General Electric (GE), which has six research centers worldwide. The centers' internal investments in its aviation plant, in Vermont, have created manufacturing capabilities.

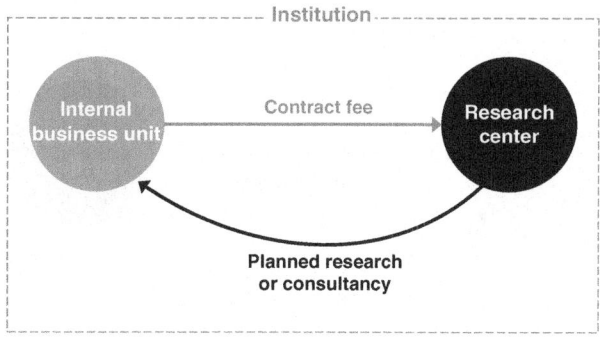

In this case, **institution** refers to a corporation or university.

Fig. 5.5 Internal contracting through transfer pricing. *Source* Prepared by the author

The Vermont team partnered with GE's Global Research Center to make specialized tools to shape advanced materials. A $75 million investment in the plant led to more than $300 million in engine production savings.

5.2.5 Freemium Product/Service

According to the European Commission, free games that feature in-app purchases account for 95–98% of the revenue in the Google Play and iTunes app stores. Only 50 companies are responsible for 81% of the highest-grossing apps. In France, the United Kingdom, Germany, and Russia, the freemium model represents 70–78% of total app revenues [17].

Is it possible to apply this model to research centers? Yes, it is.

In this model, several institutions get access free of charge to some products or services offered by the research center. In exchange, the research center might gather data from those companies, for example. The center charges additional fees if the institution wants to access premium features.

The free products or services might include a tool, an index, or a benchmark of metrics—a way to gather and share aggregated data (See Fig. 5.6).

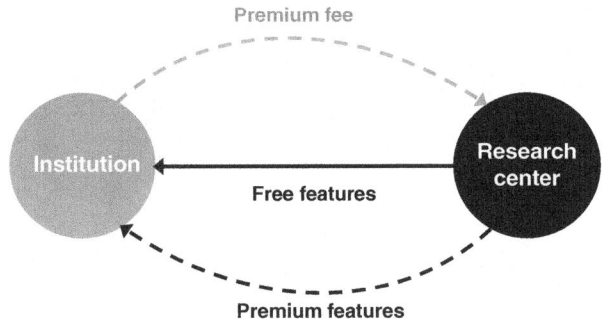

In this case, **institution** refers to a corporation or university.

Fig. 5.6 Freemium services. *Source* Prepared by the author

Noam Wasserman, the founding director of the University of Southern California's Founder Central initiative and a former professor at Harvard Business School, used this model to gather aggregated data from founding teams of start-ups. A unique data set was built up, with data about nearly 10,000 founders in the technology and life sciences industries. This data set allowed him to get the distinctive conclusions he explained in his book *The Founder's Dilemmas* [18].

Another example is the SaaSRadar proprietary benchmark database, created and collected by McKinsey's Growth Tech practice. A data set has information on 200 software-as-a-service (SaaS) companies and businesses within larger firms. This tool serves as "a diagnostic to identify opportunities for clients to grow more efficiently," in McKinsey's words.

The database covers best practices and 40 key success aspects that drive growth. It can be also used to diagnose potential pitfalls (such as in sales, marketing, and churn) [19].

5.2.6 Research Licensing

The figure of €25.6 million is quite an interesting amount to achieve from license sales during a year. Max Planck Innovation—responsible for the transfer of technology from the 83 institutes and research facilities of the Max Planck Society—predicted it would achieve this amount in a year through 80 licensing agreements [19, 20].

In this model, the research center authorizes the use or release of knowledge in exchange for a fee (See Fig. 5.7).

Every year, Max Planck Innovation evaluates 140 inventions on average, with approximately half of them leading to a patent application. Since 1979, some 4,000 inventions have been managed and around 2,300 licensing agreements concluded [21].

As an example, the software company SAP Aktiengesellschaft traditionally followed a business model that involved receiving a licensing fee up front for its software and then an annual fee of 17–18% of the original license fee for upgrades and maintenance [22].

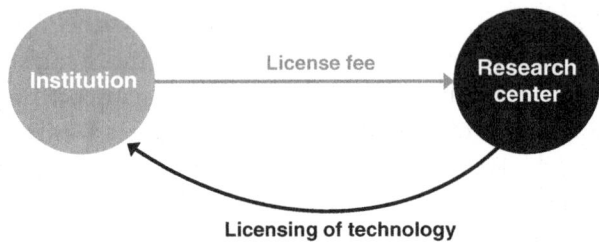

*In this case, **institution** refers to a corporation, government agency or university.*

Fig. 5.7 Research licensing. *Source* Prepared by the author

In another example, the Fraunhofer Institute for Solar Energy Systems ISE in Germany and the Italian company CGA Technologies concluded a licensing contract for a new technology—called FracTherm—which was developed and patented by Fraunhofer ISE. With the licensing contract, the research institute and the company brought two technologies together to form and commercialize one product [23].

The Chinese Academy of Sciences' Institute of Biophysics has also licensing agreements with companies, such as MicroConstants, to develop antibody drugs [24].

5.2.7 Technology Transfer by Public Funding

Many of today's innovations would not have been possible without research and development enabled by public research. Well-known contemporary examples include recombinant DNA technology, the global positioning system (GPS), MP3 technology for data storage, and voice recognition technologies such as Apple's Siri [4].

In the model of technology transfer by public funding, the research center submits an application in response to a call for public funding (e.g., for postdoctoral researchers). If the government approves the research proposal, it sends funds to the center on the condition that

5 Stage 3: Commercialization—Designing Collaborative Business ...

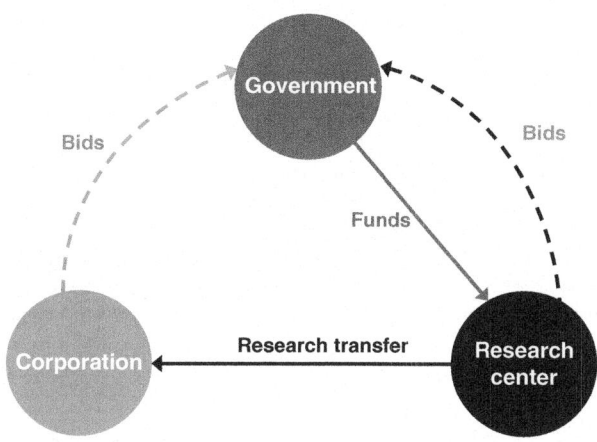

Fig. 5.8 Technology transfer by public funding. *Source* Prepared by the author

the center partner with industry to transfer knowledge to relevant companies (See Fig. 5.8).

Ideally, these companies will be more competitive, will generate more employment, and will pay more in revenue taxes to the government. Then the government will have a bigger budget to invest more in innovation, so creating a virtuous circle and empowering the whole innovation ecosystem.

The National Aeronautics and Space Administration (NASA) has documented over 1,600 such technology transfers in its Spin-off Magazine since its first edition in 1976. The commercial applications were in health and medicine, transportation, public safety, consumer goods, agriculture, environmental resources, computer technology, manufacturing, and energy conversion and use [25].

Government sponsorship of a university research center's initiatives increases the likelihood of industry involvement [26]. Therefore, it is a good way to create this virtuous circle.

The process of applying for public funds can be optimized by using several mechanisms (See Sect. 4.4). However, you first have to decide whether you want to focus on a collaborative business model based on either private or public funding, considering the advantages and disadvantages of each model (See Table 5.3).

Table 5.3 Advantages and disadvantages of public and private funding models

		PUBLIC FUNDING	PRIVATE FUNDING
PROJECT	Topic	Impact on significant groups in society – set by legislation	Emerging needs
		Difficult to sell new ideas or high-risk approaches	Easier to convince about new ideas
PROFIT	Cost of application	More bureaucracy. Complex proposals with prescribed formats and compliance procedures	Less bureaucracy. Complex proposals are not needed. More informal and willing to help with the proposal
	Income and resources	Given by one funder that has the most money	Given by multiple funders, leveraging resources
		Pays all project costs and/or indirect costs	Less likely to cover all costs and most do not cover indirect costs
		Bigger grants	Smaller grants
RISK	Bounce rate	More applicants – grants available to wider range of organizations (e.g., also nonprofit)	Fewer applicants
	Learning curve	Reviewers tend to favor established applicants	More flexibility on applicants
	Political trends	Availability of funds affected	Affected in other ways
	Flexibility on results	No. Accountable to elected officials if the rules are not followed	Yes. Priorities can change rapidly
	Available information	Easy to find information about project needs, deadlines, renewal opportunities and processes	Unwritten rules
		Closed deadline	Open deadline
		Wide range of staff	Limited staff. Fewer opportunities to contact them
		Clear reasons for rejection	May not be clear reasons for rejection

Source Adapted by the author from several sources such as Hall, M. *Getting funded: the complete guide to writing grant proposals*. (Continuing Education Press, 2003) [27]. Complemented by and refined with additional interviews and field analysis. *Note* Dark-grey cells are disadvantages, light-grey cells are neutral, and white cells are advantages

5.2.8 Creation of Spin-offs from the Research Center via External Investment

In 1972, five employees of IBM quit their jobs and started a new venture called SAP. Their idea was to develop and market standard software for business administration. Their business model combined several challenges to established industry practices of the time: standardization instead of customer-specific programming, integrated modules addressing and linking firms' multiple data needs as well as real-time instead of batch computing. Since then, SAP has grown into a global leader in business software with more than 84,000 employees and more than $22 billion in annual revenues [28].

The history of SAP is far from unique. Such corporate spin-offs have been identified as drivers of innovation and industry dynamics in a number of markets. Probably the most well-known ones, including Fairchild and Intel, have been spawned in Silicon Valley's semiconductor industry.

Spin-offs are often triggered by the parent firm's reluctance to pursue employee ideas for new products or processes [29]. Evidence from the U.S. automobile [30], laser [31], and disc drive [32] industries likewise indicates how frustrated attempts to pursue innovative opportunities at the parent firm are a major driving force of the spin-off process. As is evidenced by household names such as Ford, Intel, SAP, and Adobe [33], the spin-off process frequently leads to great firms that change the history of their industries and sometimes home regions.

In this model, after identifying a commercial application, the research center transfers the results of a research project to a new firm (the spin-off). Then the research center deals with a venture capital firm, which may involve either private investment (e.g., through business angels) or public investment (i.e., through a public offering).

This model has two variations. In the first variation, a researcher leaves the company to implement the process, dealing and sharing dividends or equity with the venture capital firm and founding the new venture. Then the former researcher may sponsor other initiatives of the center or may give an endowment (See Fig. 5.11). When leaving, the

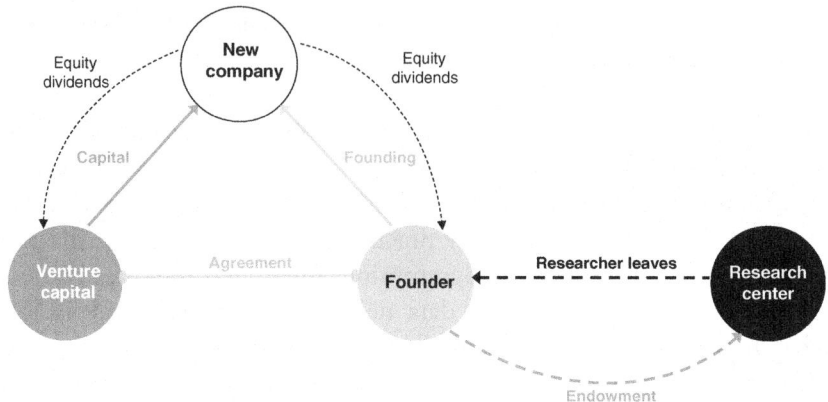

Fig. 5.9 Spin-off creation from a research center (or a researcher who leaves) via external investment. *Source* Prepared by the author

researcher ensures he or she does not use any of the intellectual property of the former employer.

In the second variation, it is the research center directly that deals with the venture capital firm, transferring knowledge to the new spin-off and sharing the dividends or equity with the venture capital firm (See Fig. 5.9).

More examples are the three new spin-offs from various Max Planck Institutes in Germany and a new shareholding, with four additional follow-up financing rounds with a total volume of more than €8 million. Since the early 1990s, 117 spin-offs have emerged from this institution, resulting in 24 merger and acquisition deals and in the creation of over 3,000 jobs [21]. Another success story is the University of Melbourne spin-off Fibrotech Therapeutics, sold for $557.5 million[7] to the Dublin pharmaceutical company Shire in 2014 [34].

You may ask what life is like for a researcher after the creation of a spin-off inside and outside the research center. Using data from the spin-offs of a Spanish university over a 10-year period, empirical results suggest that technology transfer and networking at university spin-offs decreased after their early years but, at the same time, relationships with customers improved. In other words, support from university decreases but the connection with the market increases [35].

Finally, note the importance of a knowledge transfer unit for supporting spin-offs during the process to achieve a successful transition (See Sect. 4.4). One example of knowledge transfer support is Cyclotron Road, an early-stage energy technology incubation program at Lawrence Berkeley National Laboratory in California. Launched in 2014, it aims to identify and support innovators of advanced energy technology, providing them with the tools, capital, and partners needed to commercialize their technology [36].

According to Oxford University Innovation, formerly known as Isis Innovation, these entrepreneurial or technology-transfer centers support on average up to 30 ventures and internal start-ups per year. Internal teams generally have from 2 to 12 full-time members, comprising a mixture of business professionals, entrepreneurs, technical experts, and recent alumni [37].

These supporting units have become more crucial in recent years because of a decline in venture capital investment in early-stage science companies (e.g., in life sciences), which is posing a challenge to the model [38].

5.2.9 The Search Model

The identification of technology customers—those who will use our inventions—is one of the main difficulties in commercialization. In other words, it is increasingly difficult to attract contracts for research projects [1, 39]. In conjunction with a lack of resources and even the absence of a brand—in the case of young research centers—this challenge may create a nightmare process.

An alternative model suggested by several studies is to use external networks and intermediaries [1, 39, 40]. In the search model, the center attracts contracts using an external searcher.

The external searcher is an individual—or a group of individuals, if you want to scale the model. The searchers have a lot of connections in one particular industry (e.g., a former chief marketing officer in the tech industry, a former chief technology officer in the healthcare industry, or a former consultant in the manufacturing sector). These connected individuals already know the decision-makers in those industries and have experienced the challenges of the sector. They are a perfect fit to help research centers become better connected.

The searchers, in exchange for attracting contracts, can start being paid a fee related to the size of contracts. After the performance of the searchers is validated, then they can be contracted as internal employees (See Fig. 5.10).

Very similar to this model, there is a new emerging role within research centers at corporations (especially in the pharmaceutical industry) called the scouter—a person within the institution who maps the innovation ecosystem and attracts new opportunities to the company. Companies such as Roche and Merck are already implementing this model [41] (See Fig. 5.11).

5.2.10 The Consultancy Joint Venture

In the case of a consultancy joint venture, the research center partners with a consulting firm to disseminate knowledge generated in the research center and to attract opportunities from industry. In exchange, the consultancy can leverage the research center's brand and experts. Both entities invest time and money in the project.

This model is common among young research centers that lack the internal infrastructure to execute a specific project or mature centers that want to scale the attraction of opportunities. An example of the second case is the continuous collaboration between Columbia Business School's faculty and PwC's Strategy & (the former Booz & Company) in joint dissemination initiatives [42].

Nevertheless, managing directors face the challenge of how to avoid consultancies entering into subsequent contracts with those corporations, after the end of the first collaboration with the research center. The solution? They work with a very reduced number of consultancies, with consultants that they know personally and have worked with for a long time.

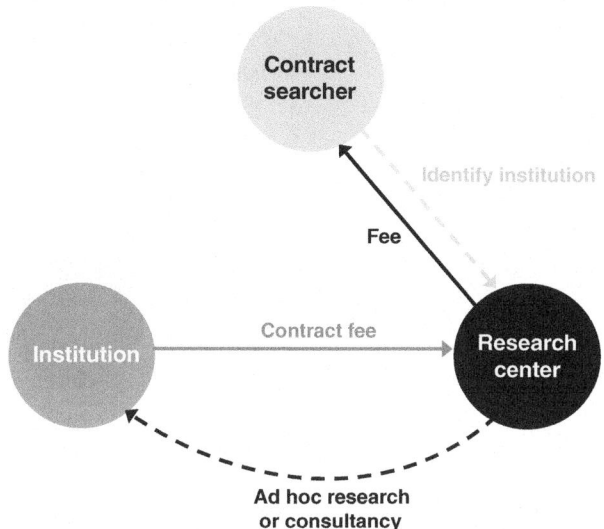

Fig. 5.10 The search model. *Source* Prepared by the author

5.2.11 Short-Term Marketing Collaboration

In the United States, the National Science Foundation's Engineering Research Centers reported in 2012 that, among their 20 associated centers, industry membership ranged from seven to 47 companies per center (averaging 23 per center). Of those companies, 48% were large firms (more than 1,000 employees) and 43% were small firms (fewer than 500 employees) [43].

This type of collaboration entails an institution that funds a project for charitable, corporate social responsibility or visibility reasons. Institutions expect the research to create specific marketing opportunities to bring their executives into conversation with potential clients, to demonstrate their corporate social purpose in ways that attract customers, and to enhance the firm's brand (See Fig. 5.12).

These projects can be of many types: a report, a book about a field of interest to potential clients of the institution, a conference or an event

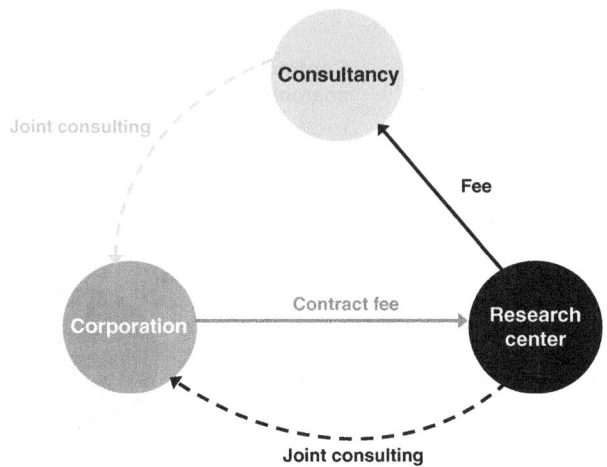

Fig. 5.11 The consultancy joint venture. *Source* Prepared by the author

at which attendees are potential clients, a business case in which the firm is the company analyzed and so gains visibility when the case is discussed in class.

A way to identify potential sponsors is first to identify their sponsoring strategy—for instance, by checking online reports, conferences, or materials that are similar to what you expect to do in your field and identifying who the industry partners are in those materials. It is easier to convince a firm that it should sponsor your research when the company has already sponsored previous projects because it has the internal processes to approve or reject similar sponsorship of other companies. It is more difficult, for instance, to attract R & D funding in industries with a poor track record for that type of funding [43].

However, it is not easy to identify the person whose role is to make those decisions in a potential partner. The person responsible might be, depending on the topic and the material, the chief marketing officer, the chief innovation officer, the sponsorship director, the partnership director, etc.

Marketing techniques for this purpose include publishing newsletters, visits to industry by directors and faculty, visits to the research center by industry representatives, exhibits at trade association

5 Stage 3: Commercialization—Designing Collaborative Business …

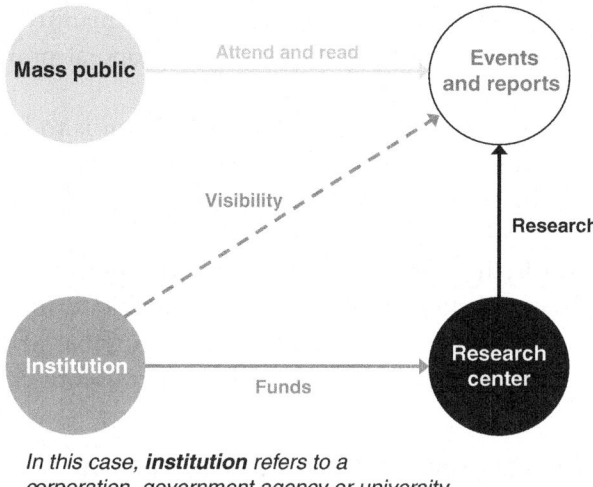

In this case, **institution** refers to a corporation, government agency or university.

Fig. 5.12 Short-term marketing collaboration. *Source* Prepared by the author

meetings, participation in technical society conferences, publication of white papers, participation in industry research consortia, a website, videos, workshops, meetings with alumni in the industry, student mentoring by industry, extended visits to the center by industry researchers, industrial advisory boards, hosting center tours for members and their clients or prospective clients, and short courses.

Four of these initiatives were selected as very impactful methods to identify prospective industrial partners for research centers, according to center managing directors: leads from faculty, personal contacts at trade meetings, publishing, and having quality facilities [44].

Of the benefits to institutions, five were selected as being of the most value: access to new ideas or technology, access to the center's faculty, networking with other industry members, access to facilities, and employing the center's graduates. The longer a company has been a member, the more likely it is the company will continue [44].

Centers should target specific institutions based on their involvement in the particular industry, their interactions with other sponsors, and their degree of involvement in technology development [43].

Finally, it is recommended to specify or even quantify those value propositions or features that would make the proposal more attractive to the target institution's decision-makers.

Imagine, for example, that your value proposition is to host and disseminate one conference a month, to show your company's logo in the brochures and the slides of the presentation, to include your name in the press release, etc.

In other words, it is important to translate intangible assets (e.g., knowledge, data, and skills) into tangible measurable features of a value proposition (e.g., number of media appearances, number of attendees at a conference, number of publications, and number of jobs created).

5.2.12 Long-Term Marketing Collaboration

Faculty chairs at Duke University in North Carolina can be established for between $1 million and $5 million depending on the sponsored profile, from a visiting professor to the dean. Similar ranges have been adopted in research centers [45].

These chairs are just one example of long-term marketing collaboration, in which a firm has more general marketing and corporate social responsibility incentives, incrementally building a brand with elite or general audiences that associate the firm with attractive or socially worthwhile research, as if the company were supporting a symphony orchestra or a medical charity (See Fig. 5.13).

University chairs can be funded by individuals or corporations. Depending on the type of endowment, the researcher receives a yearly budget to invest in research projects relating to a specific field, as defined by the chair.

Other examples include the Daniel L. Alspach chair of dynamic systems and controls at the University of California San Diego, the European Union Jean Monnet chair at the New York University School of Law, and the David Mulvane Ehrsam and Edward Curtis Franklin chair of chemistry at the Stanford University School of Humanities and Sciences.

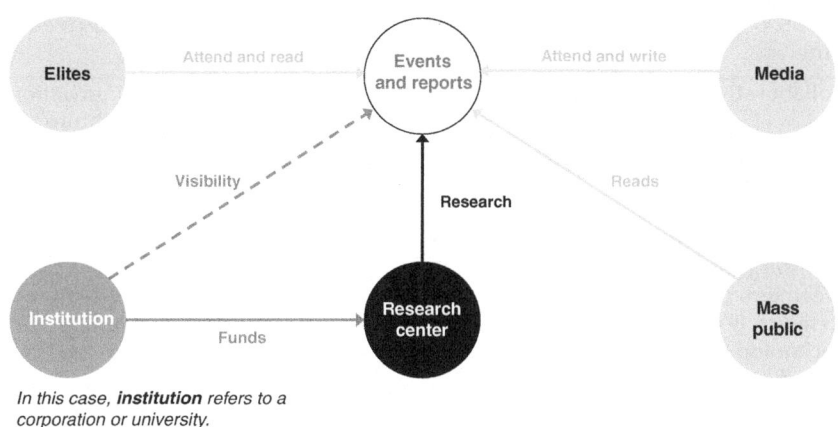

Fig. 5.13 Long-term marketing collaboration. *Source* Prepared by the author

5.3 Select the Collaborative Business Model that Fits the Orientation and Age of Your Center

If you compare the income statement of different research centers, you will find different strategies to attract revenues. For instance, the Fraunhofer-Gesellschaft (Fraunhofer Society)—Europe's largest innovation-oriented research organization—received contract revenues mainly from industry (€641 million)[8] and the public sector[9] (€10 million from contracts and €531.6 million from project funding) out of its €1.36 billion total annual research revenues for 2012 [46, 47].[10]

On the other hand, the University of Toronto's Faculty of Applied Science and Engineering earned research revenues mainly from sponsors (C$81.6 million in 2015) and from industry (C$7.7 million) [48]. In other words, industry funding was 40% in the first case and 8.6% in the second case.

Based on the results of interviews with research center leaders and the analysis of 3,881 research centers, 12 collaborative business models are proposed depending on the type of center, specifically its orientation and age (See Table 5.4).

Table 5.4 Collaborative business model recommended for each type of research center

Collaborative business model	Research young	Research mature	Innovation young	Innovation mature
Short-term external contracting		x		x
Medium-term external contracting		x		x
Long-term external contracting		x		x
Transfer pricing		x		x
Freemium product/service	x	x	x	x
Research licensing		x		x
Technology transfer by public funding	x	x	x	x
Creation of a spin-off via external investment		x		x
Search model	x		x	x
Consultancy joint venture		x		x
Short-term marketing collaboration		x		x
Long-term marketing collaboration		x		x

Source Prepared by the author

5.4 Partner with Complementary Brands and Write Media Reports

Do you lack a recognizable brand or experienced researchers? Are you unable to attract industry partners for collaborations? Was your center founded recently? Are you consolidated but want to extend your institution's brand?

A quick way to establish a brand is to use brand architecture in which you leverage the brand of already positioned institutions or companies [49].

Collaborative research branding initiatives, which are mutually beneficial for the parties involved, may be established. You can collaborate with those brands to leverage research capability or dissemination capability.

In the first case, you leverage an external stakeholder regarding the costliest aspect of a publication: the creation of results. The costs of

5 Stage 3: Commercialization—Designing Collaborative Business ...

research production include the costs of researchers carrying out field and lab research activities and consolidating and writing up the results of their research. These costs account for 66% of the total cost. Reading, which includes the cost of reading by researchers in institutions,[11] accounts for 19% of the total cost (See Fig. 5.16). The rest of the total goes on other costs such as publishing and distribution, access provision, user research, and price cost [50].

This practice is similar to the collaborative business seen in previous sections from the research institution perspective (See Sect. 5.2.10 and Fig. 5.14).

In the second case, you externally leverage the dissemination capability of a brand. A research center can extend its brand by partnering with firms (e.g., consultancies) that disseminate their knowledge and content. These partnerships may offer opportunities to increase the impact of brand awareness in other geographic areas, increase brand perceptions, and give the knowledge created greater outreach.

The innovation consultancy Opinno, for instance, uses a partnership to disseminate in Latin America and Spain the annual ranking of the top young innovators of the Massachusetts Institute of Technology under the name Innovators Under 35 [51] and it also publishes the *Harvard Business School Review* in Spanish [52].

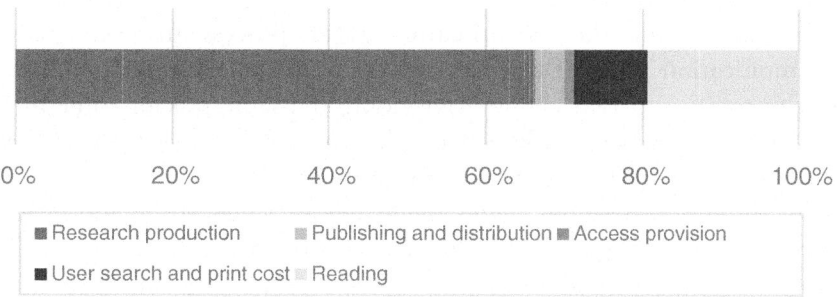

Fig. 5.14 Percentage of the annual cost incurred in the global scholarly communication process, by value chain component. *Source* Prepared by the author using data from *Activities, costs, and funding flows in the scholarly communications system in the UK* (Research Information Network, May 2008)[12]

5.5 Review the Communication Processes in Your Center: Map of Functions, Cascade Processes and Customer Relationship Management of Specialist Media

Are you creating more value (e.g., knowledge, publications, and conferences) than the value perceived by the nonacademic market? Are your researchers unaware of who to contact within your organization to maximize the impact and outreach of their research results?

Research institutions usually follow complex matrix models of organizational structure. However, these organizations are usually segmented in silos with a high density of corporate politics—not an easy scenario for internal communications [53].

In high-performance research centers, three principles were identified to optimize the propagation of results.

First, fill out a map of functions. This is not a who's who or a map of the research center by business unit but who I have to contact when I have to do something. For example, who should I contact when I want to have something disseminated in the international press? Who should I contact if I want to give a lecture about the results of the topic? Who should I contact if I want to reach a company in the relevant industry to apply the results obtained from the research project? and so on. Then give this map to each researcher.

Second, redefine the dissemination cascade process. In several cases, communication units of research centers were somehow unconnected. As a consequence, two people were doing the same task or something very similar, they were talking with the same external contact or they were using only a few channels to communicate.

For instance, a research center might release a new study and give it visibility only through social networks but not in the press. Or in the press but not in internal channels. Or in an internal magazine but not on the website. However, a cascade process leverages your internal structure to maximize the external impact. Centers such as those at University of Michigan are already applying these principles, getting best practices from the tool kit that the university shares among research units in this and other aspects.

One example of communication in cascade would be:

1. The researcher tells the person who centralizes the communication channels that a new study has been released.
2. A person in the communication unit translates the research results into insights that will be understandable to readers who are not experts in the field.
3. A person transforms that document into a press release and shares the document with the media.
4. A person transforms that press release into a website article.
5. A person transforms that website article into a blog article.
6. A person transforms that blog article into a post on social networks.
7. A person transforms the study into a lecture.
8. A person transforms the lecture into an event on the website.
9. A person transforms the event on the website into a post on social networks.

And so it would continue.

Third, the rejection rate for articles in the established media is high. One way to increase the acceptance rate is to give specific pieces of information to a very segmented type of journalist based on those journalists' interests. To do this, in addition to creating long-term relationships with those journalists, it is important to generate internal customer relationship management (CRM) with all the journalists, segmented by topic of interest, geographic area, and medium. With this tool, when the communication unit receives a new study, it will know to whom specifically it should write. This is similar to the corporate relations office at Duke University, which coordinates interactions with industry.

5.6 Partner with Visiting Researchers or Reward Recognized Faculty

Is your center too young to attract funding or ad hoc research contracts? Do you lack internal research capability?

In this case, you could partner with researchers who are renowned in their field for a specific collaboration (e.g., the submission of a proposal to obtain public funds) or for long-term collaborations (e.g., to create long-term research projects, or a series of lectures). Additionally, you can leverage external brands by creating awards for top researchers.

The IBM Faculty Awards, for example, support basic research, curriculum innovation, and educational assistance in specific focus areas. The program is intended, first, to foster collaboration between researchers at leading universities worldwide and those in IBM research, development, and services organizations; second, to promote curriculum innovation to stimulate growth in disciplines and geographic areas that are strategic to IBM.

The awards are not contracts but cash awards granted annually. The current maximum award for any one recipient is $40,000 per year. Intellectual property rights are not specified [54].

5.7 Give Periodic Lectures to Industry, Translating Research Results into Quantified Impact

Does the market fail to understand what you are doing in your research center? Do you lack a network for university-industry collaborations? Are you unaware of how to validate the market impact or the market opportunity of your research project? Do you have a discovery that is difficult to understand? Do you have a technology the generated value of which your customers fail to perceive? Do the evangelists of your discoveries find it difficult to explain why your discovery should be applied in their companies?

A mechanism that research centers' transfer offices are implementing to resolve the problems mentioned is periodic lectures to industry leaders. In these lectures, either professors with experience in industry–university collaboration or the director of the technology-transfer unit would explain the implications and applications of the research projects, explaining the value propositions via quantified value and success cases.

The technology transfer unit at the Barcelona Supercomputing Center has monthly meetings, by sector, to explain the implications of its discoveries [55].

The French National Center for Scientific Research, each year, gives access to its laboratories and technological platforms for training sessions in specific areas. The center has achieved 4,535 patents and 1,237 active licenses [56].

Similar mechanisms include a series of industry speakers. Industry professionals would be invited to the center for speaking engagements, followed by meetings with the center's faculty and a look at the center's facilities. The alumni office could help with introductions to key stakeholders in industry. Attendees of the conferences appreciate being given opportunities to network with other industry members during or after the speaking sessions [57].

From a sales and marketing perspective, intangible assets—such as knowledge—are difficult to sell. However, you can use the common language of potential buyers (e.g., how the application of your discovery has increased the bottom line of other companies, how it has decreased their costs by optimizing their processes, how it is changing the rules of the sector).

This explanation can be directly done verbally or via publications. Industry leaders can read about the insights in greater detail in publications (e.g., articles in the press), thus developing an interest in what you are doing.

MIT Technology Review is an example of a magazine that shares insights from faculty—and other experts—to equip its audiences with the intelligence to understand a world shaped by technology.

A similar case is that of the 400 *Perspectives* indexed articles, periodic and concise essays released by the consultancy The Boston Consulting Group since 1964, designed to simulate senior management thinking on a range of business issues. The consultancy's founder Bruce Henderson referred to these insights as "a punch between the eyes" [58].

5.8 Adapt Team Size to Market Needs

Do you have a recognized and renowned research center but lack industry collaborations? Have you talked already with all the big players in the sector in which you specialize and did not know how to keep your business-to-business (B2B) customer portfolio growing? Do you frequently talk with business professionals rather than research professionals as potential clients?

The internal architecture of research centers can create a huge barrier for commercialization. For example, although Spain has three centers among the top 200 worldwide, 70% of the research projects at university centers are sold outside the country [6].

Why? In Spain, 99.9% of local industry is composed of small and medium-sized enterprises (SMEs). The budget, expertise, and internal structures of these companies are unable to absorb large research teams. For instance, SMEs might not have a large enough budget or they might lack the internal knowledge to talk with the research center's experts or absorb the center's discoveries.

In line with this example, data analysis of academic-industry collaboration across 125 Spanish research centers (See Fig. 5.15) illustrates how the size of a center's research team affects its research output and how the economic impact (funding) is related to the academic impact (publications in peer-reviewed journals).

The chart illustrates how the size of research team per project (horizontal axis) is related to the total funding of current projects and the number of indexed papers in journals during a year (vertical axis).

The analysis shows how both metrics (i.e., project funding and indexed papers) are maximized when the size of the research team is between two and three researchers. Additionally, teams of fewer than seven employees generally perform better at both goals: publishing in indexed journals and funding research projects.

In conclusion, it is important to consider the relevant industry's characteristics when designing the internal structure of the research center and the research teams.

Motivating faculty members to take opportunities and to interact with industry and multi-institution coordination are two of the challenging aspects of being a managing director at a research center (See Table 1.1). Barriers are generally caused by the center's internal politics and bureaucracy.

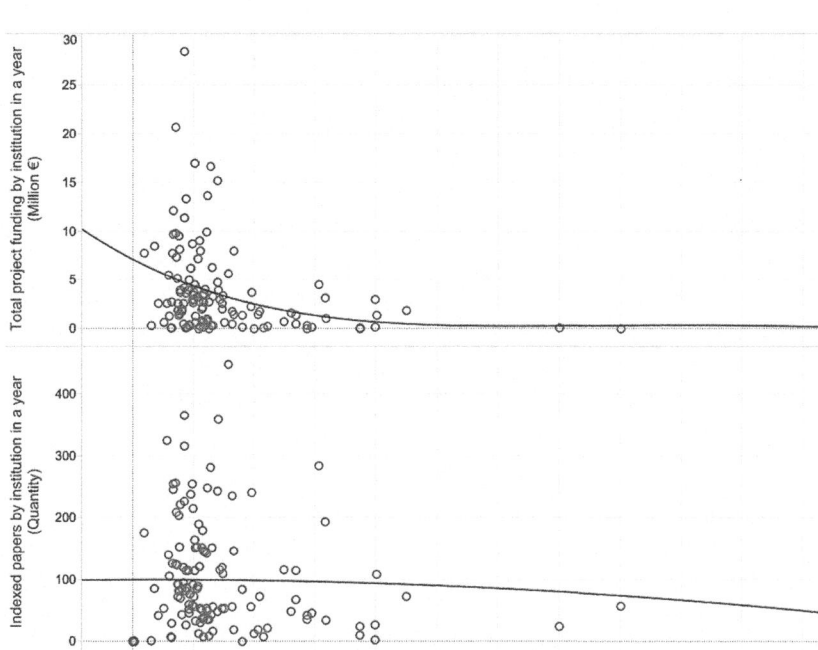

Fig. 5.15 Comparison: size of research teams with total funding (*top*) and indexed papers (*bottom*). *Source* Prepared by the author with Tableau software using data adapted from 125 Spanish research centers. The initial data were collected by the Spanish National Research Council (CSIC) [46]. *Note* The size of research team per project was calculated as the sum of total employees—including scientific, management, and external employees—divided by the number of current projects at the selected research center. The *line* in each chart represents the exponential trend line

5.9 Identify the Decision Makers and Their Key Performance Indicators

Are you designing a strategic plan but you do not know how to take into account the vision and mission of your research center? Have you secured a commercialization opportunity but do not know which researcher is the most appropriate (in terms of both area of expertise and interest in the collaboration)? Are you going to implement a new commercialization initiative but you do not know which stakeholders in your institution should approve the proposal or whether you should ensure their buy-in?

Several authors have pinpointed the importance of identifying and mapping those "unwritten rules" in your center to advance and leverage them [59]. Centers are applying this principle. In 2003, the pharmaceutical company Merck introduced a scouting organization within the Word Wide Licensing and Knowledge Management group, growing from 11 to 65 employees in 2011. It is a team that generates novel opportunities for the company, developing connections with Merck's internal research units and with outside partners such as entrepreneurs and venture capital firms. Scouters has a tough role connecting researchers with entrepreneurs and investors, who have different performance indicators [41]. Additionally, scouters have to convince internal teams and decision-makers to see fructifying those opportunities.

These problems can be solved by gaining an understanding of the organization. However, as we have seen, research centers are complicated organizations. To make things easier to understand, I suggest relying on three considerations: interests, performance indicators, and the role in the pyramid. First, the research map (See Sect. 3.3) is used to identify their research interests. Second, the preferences and key performance indicators (KPIs) of the different stakeholders are identified. Third, the research/faculty pyramid is used to identify the characteristics of their roles. All of these considerations can be summarized to understand the mind-set of the researcher or executive with whom you are sitting.

Since each business unit's director has different KPIs to be measured in the quarterly results, although these are often unwritten, you should identify what those KPIs, those unwritten rules, are [59].

The first differentiation is between academic and nonacademic roles. Academic roles may have KPIs related to academic indicators (e.g., number of published articles in top peer-reviewed journals, number of invested hours, visibility among other academics in their field

5 Stage 3: Commercialization—Designing Collaborative Business ...

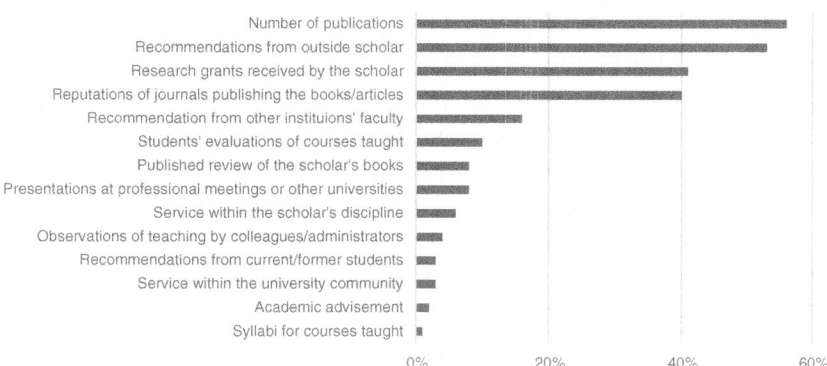

Fig. 5.16 Aspects that faculty rated as important for granting tenure. *Source* Prepared by the author using data from Boyer [60]

of expertise, and network access to senior editors in top peer-reviewed journals). In short, publishing, reputation, and time (See Fig. 5.16).

Executive roles may have economic indicators (e.g., profitability of the initiative, number of invested hours, number of media appearances, and number of paid attendees at a conference).

The second differentiation is to keep in mind the seniority of the researcher. While young researchers/faculty have more focused priorities, senior researchers/faculty may have more diverse indicators, especially among those who have achieved tenure.

Those who are more senior may have less pressure to be published in peer-reviewed journals, a higher teaching workload, and new interests in other initiatives (e.g., internal and external consulting, and institutional leadership).

To understand further the mind-set of those profiles (researchers and executives), it is interesting to know a few items of data. Executives in research institutions are under the pressure of having really lean budgets. However, researchers have a tough up-or-out policy, after completing the Ph.D. in over 6 years.

Depending on the sector and the geography, only 12.8% of Ph.D. graduates can attain academic positions because the reproduction rate in academia is very high. A professor graduates 7.8 new Ph.Ds. during her or his whole career on average. Only one of these graduates can replace the professor's position [61].

Additionally, they need over 7 years to achieve tenure or over 12 years to be a full professor—in the case of university [61–63]. One of the factors that can speed up this journey is the speed of publication. Nevertheless, the publication cycle—from submitting to acceptance—of a paper in a top peer-reviewed journal can be something between 100 and 150 days, in addition to the time needed to get the results and write the piece. The total time would be between 2 and 6 years (See Sect. 4.3).

Finally, having a network map of your institution and knowing the key decision-makers for your endeavors and their KPIs could increase the chances of being approved because then you could focus your explanation of your initiatives to them on the performance metrics and interests they value the most [64] (See Sect. 6.4).

5.10 Define Delivery Requirements Before Starting and Presell Your Results

Have you delivered the results of an ad hoc research project but it has not been approved? Have you been involved in a research project with a very undefined end date? Do you usually exceed the initial budget of ad hoc research projects?

In 2004, the automotive company Audi proposed a strategic collaboration with the Technical University of Munich (TUM), through the establishment of a research institute near Audi headquarters that would support more than 100 Ph.D. students working on technology and innovation issues vital to Audi's competiveness.

It is common for industry–university collaborations to have misaligned preferences. While a university has long release cycles, industry has short ones. While a university has academic preferences, industry has economic ones. While a university usually produces theoretical results, industry prefers applicable results (See Fig. 5.2).

There are two forms of best practice to align industry–university collaboration. First, requirements in terms of deliveries, timing, scope, etc. should be detailed to avoid any misalignments.

Second, the results of your ad hoc research should be presold before the final delivery, ensuring the company's buy-in. Any misalignment of expectations in the results can trigger the end of the relationship.

As Peter Tropschuh—Audi's head of scientific relations and corporate responsibility—explained, two kinds of best practice were learned from Audi's partnership with TUM. First, it is necessary to "define a clear strategy and listen": research centers should listen to industry and ask about its real needs. Second, both sides should "meet and talk regularly" about the bad and the good things, maintaining personal contact [65].

5.11 Create a Specific Unit to Apply for Public Funds or Leverage External Consultancies

Do you lack the experience and in-depth expertise to apply for public funds? Do you have a very low acceptance rate of proposals for public funding?

During the whole process of attracting and executing a project, the differences between the public and private sectors are quite significant (See Fig. 5.17).

Additionally, the knowledge required to apply for proposals is considerable. The acceptance rate for European public funds (e.g., the Horizon 2020 program) is 11.8%. This rate means that, on average, you have to send around nine proposals to get one funded. These proposals need on average between 6 and 9 months to prepare or 2–3 months working full-time [66].[13] So it is important to increase the acceptance rate somehow.

In this process, I suggest having a specialist unit of a few experts working on those funding programs, who can provide templates, benchmarks, proposals previously submitted by the institution, etc.

If your center cannot afford an additional unit, at least in the beginning, young research centers could partner with external consultancies

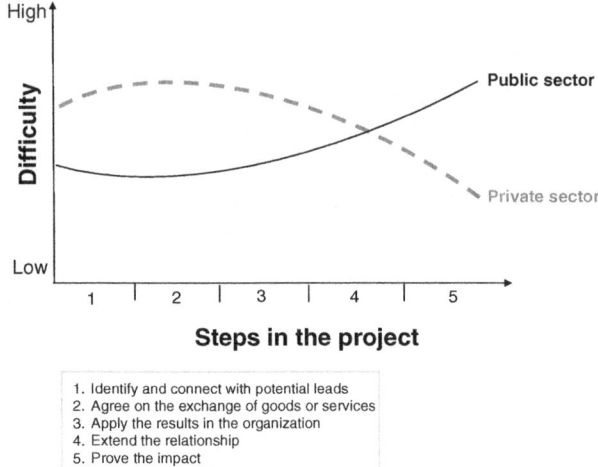

Fig. 5.17 Differences in difficulty level between public and private funding. *Source* Prepared by the author using data from several sources such as Tinkler et al. [9]

specializing in the topic. For instance, the consultancy Ateknea accelerates the delivery of proposals and tries to improve the acceptance rate. It claims it improves the acceptance rate from the initial 11.8% to at least 40% [67].

Another example is Max Planck Innovation. To validate the industrial compatibility of inventions resulting from basic research and to achieve closer links with industry and the market, Max Planck Innovation has set up various incubators such as the IT Inkubator in Saarbrücken, Germany. Saarland University, and Max Planck Innovation founded the company "to take up promising IT solutions" developed at their research facilities [20].

Endnotes

1. The estimated number is based on the calculation that there are about 25,000 universities in the world and they offer seven Ph.D. subjects, with the assumption that more than 100 people have done each Ph.D.

on average since the start of the university offering this qualification. Although some people may hold two or more Ph.Ds., we can round the total number down from 17.5 million to around 15 million. This number represents around 0.2% of an approximate global population of 7.5 billion [68].
2. Between 1998 and 2006, the number of doctorates handed out in all OECD countries grew by 40%, compared with 22% for the United States. Ph.D. production sped up most dramatically in Mexico, Portugal, Italy, and Slovakia. Even Japan, where the number of young people is shrinking, churned out about 46% more Ph.Ds. Part of the growth reflects the expansion of university education outside the United States [69].
3. The European Union reported that, in the first 2 years, Horizon 2020 allocated more than €15.9 billion in EU funding, spread over 9,000 grants.
4. Governments allocate basic funding, a budget for research institutions that are qualified under the criteria that are defined by the government in consultation between the research council, the research institution and the allocating ministry. This long-term funding aims to improve the core activities of research institutions within strategic areas.
5. The frequency of four collaborative business models was estimated based on the interviews, public information, and other sources. These four models are transfer pricing, freemium product or service, search model, and consultancy joint venture. The frequency of the other models was estimated based on a literature review.
6. In taxation and accounting, transfer pricing refers to the rules and methods for pricing transactions between enterprises under common ownership or control.
7. The amount includes initial and following payments, based on milestones.
8. Excluding €117 million in license-fee revenues.
9. Germany's federal government and the Länder.
10. The other €181 million came from research funding organizations and other sources.
11. There was some debate about whether reading should be considered part of research production or part of consumption activities. The latter approach was chosen in order to distinguish between the input and output components of research activity.

12. The estimate relies on detailed modeling of global publication and distribution costs. However, the access costs in the model are only for the United Kingdom so the global access cost estimate is an extrapolation from United Kingdom calculations.
13. The average time spent by a coordinator is 350–450 hours, which is 45–60 working days (full-time). The average time spent by a work package leader is 70–100 hours, which is 9–14 working days (full-time).

References

1. Lichtenthaler, U. External commercialization of knowledge: review and research agenda. *International Journal of Management Reviews* **7**, 231–255 (2005).
2. Florea, N. V. *Using Branding to Attract, Recruit, and Retain Talented Staff* (Editura Universitaria Craiova, 2011).
3. Rohrbeck, R. & Arnold, H. M. Making University–Industry Collaboration Work—A Case Study on the Deutsche Telekom Laboratories Contrasted with Findings in Literature. In *ISPIM Annual Conference: Networks for Innovation* (2007).
4. OECD. *Science, Technology and Innovation Outlook 2016* (OECD, 2016).
5. European Commission. *Horizon 2020—Two Years on Research and Innovation* (European Commission, 2016).
6. Gulbrandsen, M. & Smeby, J. C. Industry funding and university professors' research performance. *Research Policy* **34**, 932–950 (2005).
7. Zott, C. & Amit, R. Business model design: an activity system perspective. *Long Range Planning* **43**, 216–226 (2010).
8. Schofield, T. Critical success factors for knowledge transfer collaborations between university and industry. *Journal of Research Administration* **44**, 38–56 (2013).
9. Tinkler, J., Dunleavy, P. & Bastow, S. *The Impact of the Social Sciences: How Academics and Their Research Make a Difference* (SAGE, 2014).

10. Perkmann, M. & Salter, A. How to create productive partnerships with universities. *MIT Sloan Management Review*. Available at: http://sloanreview.mit.edu/article/how-to-create-productive-partnerships-with-universities/ (2012). (Accessed 7 Mar 2017).
11. Cascio, W. F. Evidence-based management and the marketplace for ideas. *Academy of Management Journal* **50**, 1009–1012 (2007).
12. Pertuzé, J. A., Calder, E. S., Greitzer, E. M. & Lucas, W. A. Best practices for industry–university collaboration. *MIT Sloan Management Review* **51**, 83–90 (2010).
13. Barragán-Ocaña, A. & Zubieta-Garcia, J. Critical factors toward successful R & D projects in public research centers: A primer. *Journal of Applied Research and Technology* **11**, 866–875 (2013).
14. *Measuring the Digital Economy* (OECD Publishing, 2014).
15. PwC. *PwC and Google Announce Joint Business Relationship*. Available at: http://press.pwc.com/News-releases/pwc-and-google-announce-joint-business-relationship/s/091c203d-1de7-466c-8574-f0ac96593edc (2014). (Accessed 1 Mar 2017).
16. Imperial College London. *Shell-Imperial College Grand Challenge on Clean Fossil Fuels*. Available at: https://www.imperial.ac.uk/energy-futures-lab/research/our-projects/shell-imperial-college-grand-challenge-on-clean-fossil-fuels/. (Accessed 7 Mar 2017).
17. Probst, L., Frideres, L., Pedersen, B., Lidé, S. & Kasselstrand, E. *New Business Models Freemium: Zero Marginal Cost* (European Commission, 2015).
18. Wasserman, N. *The Founder's Dilemmas. The Kauffman Foundation Series on Innovation and Entrepreneurship* (Princeton University Press, 2012).
19. McKinsey & Company. *Growth Tech*. Available at: http://www.mckinsey.com/industries/high-tech/how-we-help-clients/growth-tech. (Accessed 1 Mar 2017).
20. Max-Planck-Gesellschaft. *MPG Annual Report 2014* (2015).
21. Max Planck Gesellschaft. *Annual Report 2015* (2016).
22. McGrath, R. G. & Macmillan, I. C. How to rethink your business during uncertainty. *MIT Sloan Management Review* (2009).

23. Fraunhofer Gesellschaft. *Fraunhofer ISE and CGA Technologies S.p.A. Conclude Licensing Contract—FracTherm® Technology in Roll-Bond Solar Absorbers*. Available at: https://www.ise.fraunhofer.de/en/press-media/press-releases/2013/fraunhofer-ise-and-cga-technologies-spa-conlude-licensing-contract.html (2013). (Accessed 2 Mar 2017).
24. Chinese Academy of Sciences. *MicroConstants Leverages Chinese Academy of Sciences Antibody Rights in License Deal* (Chinese Academy of Sciences, 2015).
25. National Aeronautics and Space Administration. *Spinoff* (2017).
26. Craig Boardman, P. & Ponomariov, B. L. University researchers working with private companies. *Technovation* **29**, 142–153 (2009).
27. Hall, M. S. & Howlett, S. *Getting Funded: The Complete Guide to Writing Grant Proposals* (Continuing Education Press; Portland State University, 2003).
28. SAP. *SAP Quarterly Statement—Preliminary Q4 and FY 2016 Results* (2016).
29. Klepper, S. Silicon valley—A chip off the old Detroit bloc. In *Entrepreneurship, Growth, and Public Policy* 79–115 (2009).
30. Klepper, S. The capabilities of new firms and the evolution of the US automobile industry. *Industrial and Corporate Change* **11**, 645–666 (2002).
31. Klepper, S. & Sleeper, S. Entry by spinoffs. *Management Science* **51**, 1291–1306 (2005).
32. Christensen, C. M. the rigid disk drive industry: A history of commercial and technological turbulence. *Business History Review* **67**, 531–588 (1993).
33. Chesbrough, H. The governance and performance of Xerox's technology spin-off companies. *Research Policy* **32**, 403–421 (2003).
34. Science Meets Business. *Top 25 R & D Spin-offs* (2016).
35. Pérez Pérez, M. & Sánchez, A. M. The development of university spin-offs: Early dynamics of technology transfer and networking. *Technovation* **23**, 823–831 (2003).
36. Lawrence Berkeley National Laboratory. *Cyclotron Road—Mission*. Available at: http://www.cyclotronroad.org/mission/. (Accessed 28 Feb 2017).
37. Oxford University Innovation. *How to Set Up a Successful University Startup Incubator* (Oxford University Innovation, 2014).

38. Fleming, J. J. The decline of venture capital investment in early-stage life sciences poses a challenge to continued innovation. *Health Affairs* **34**, 271–276 (2015).
39. Elton, J. J., Shah, B. R. & Voyzey, J. N. Intellectual property: Partnering for profit. *McKinsey Quarterly* 58–67 (2002).
40. Secchi, E. in *Agent-Based Simulation of Organizational Behavior* 329–344 (Springer, 2016).
41. Monteiro, F. & Klueter, T. M. *Building the Virtual Lab: Global Licensing and Partnering at Merck* (INSEAD, 2014).
42. Strategy&. *Rita Gunther McGrath on the End of Competitive Advantage* (*Strategy&*, 2014). Available at: https://www.strategy-business.com/article/00239?gko=ede47. (Accessed 10 Apr 2017).
43. Sander, E. *ERC Best Practices Manual Chapter 5 Industrial Collaboration and Innovation* (National Science Foundation, 2013).
44. National Science Foundation. *Engineering Research Centers Best Practices Manual. A Collaborative Effort of the NSF Engineering Research Centers* (National Science Foundation, 1997).
45. Duke University. *Endowment Giving*. Available at: https://dukeforward.duke.edu/ways-to-give/endowment/endowment-giving/. (Accessed 10 Mar 2017).
46. Fraunhofer Gesellschaft. *Annual Report 2012* (2012).
47. Fraunhofer Gesellschaft. *Annual Report 2015* (2016).
48. University of Toronto's Faculty of Applied Science and Engineering. *Annual Report: Performance Indicators* (2015).
49. Uggla, H. The brand association base: A conceptual model for strategically leveraging partner brand equity. *Journal of Brand Management* **12**, 105–123 (2004).
50. Pontes, B. & Frases, S. The Cryptococcus neoformans capsule: Lessons from the use of optical tweezers and other biophysical tools. *Frontiers in Microbiology* **6**, 1–24 (2015).
51. MIT Technology Review. *Innovators Under 35 | Europe*. Available at: http://www.innovatorsunder35.com/europe. (Accessed 2 Mar 2017).
52. Opinno. *Harvard Business Review*. Available at: http://www.opinno.com/es/users/harvard-business-review. (Accessed: 2nd March 2017).

53. Alpert, D. Performance and paralysis: The organizational context of the American Research University. *The Journal of Higher Education* **56**, 241 (1985).
54. IBM. *IBM University Research and Collaboration.* Available at: http://www.research.ibm.com/university/awards/faculty_innovation.shtml. (Accessed 7 Mar 2017).
55. Barcelona Supercomputing Center. *Technology transfer | BSC-CNS.* Available at: https://www.bsc.es/innovation-and-services/innovation-for-industry/technology-transfer. (Accessed 2 Mar 2017).
56. The National Center for Scientific Research. *Innovation and Business Relations.* Available at: http://www.cnrs.fr/en/aboutcnrs/innovation-business.htm. (Accessed 10 Apr 2017).
57. Boschi, F. C. *Best Practices for Building and Maintaining University–Industry Research Partnerships: A Case Study of Two National Science Foundation Engineering Research Centers* (Montana State University, 2005).
58. Stern, C. W. & Deimler, M. S. *The Boston Consulting Group on Strategy: Classic Concepts and New Perspectives* (Wiley, 2012).
59. Beeson, J. *The Unwritten Rules: The Six Skills You Need to Get Promoted to the Executive Level* (Jossey-Bass, 2010).
60. Boyer, E. L. *Scholarship Reconsidered: Priorities of the Professoriate* (Carnegie Foundation for the Advancement of Teaching, 1990).
61. Larson, R. C., Ghaffarzadegan, N. & Xue, Y. Too many Ph.D. graduates or too few academic job openings: the basic reproductive number R0 in academia. *Systems Research and Behavioral Science* **31**, 745–750 (2014).
62. Harvard University. *Appointment and Promotion Handbook* (Harvard University, 2014).
63. National Science Foundation. *Science and Engineering Indicators* (National Science Foundation, 2012).
64. Miller, R. B., Heiman, S. E. & Tuleja, T. *The New Conceptual Selling: The Most Effective and Proven Method for Face-to-Face Sales Planning* (Warner Business Books, 2005).
65. Edmondson, G., Valigra, L., Kenward, M., Hudson, R. L. & Belfield, H. *Making Industry–University Partnerships Work: Lessons from Successful Collaborations* 1–52 (Business Innovation Board AISBL, 2012).
66. Petrović, M. *How to Prepare Project Proposals for HORIZON 2020* (University of Zagreb, Faculty of Political Sciences, 2014).

67. Ateknea. SME Instrument: Ateknea Continues Strong Record. Available at: http://ateknea.com/news/ateknea-leadership-smes/ (2015). (Accessed 9 Mar 2017).
68. Worldometers. Current world p opulation. Available at: http://www.worldometers.info/world-population/. (Accessed 1 Mar 2017).
69. OECD. Reviews of innovation policy France. *Innovation*, 269 (2007).

6

All Stages: Innovation Ecosystem—Qualifying and Leveraging the Internal and External Agents Based on Merit

Abstract This chapter pinpoints the five causes behind not leveraging appropriately a research center's innovation ecosystem. These are a lack of understanding of the innovation ecosystem, internal gaps, no external proximity, a lack of internal resources or hooks to keep talent, and few interactions among the ecosystem's agents. Nine practical mechanisms being applied by recognized centers to confront these issues are then presented: qualifying the stakeholders of your innovation ecosystem; connecting with your internal decision makers, influencers and advisors; adapting your commercialization model to the characteristics of your ecosystem; connecting virtually with disperse agents; crowdsourcing the areas of your value chain that are not in the core business; capitalizing on aging; moving from academics to entrepreneurial academics; and recognizing academic entrepreneurs before they leave.

Keywords Innovation ecosystem · Entrepreneurial University · Academic entrepreneur · Entrepreneurial scientist · University-industry-government · Academic intrapreneurship · Crowdsourcing · Aging capitalization · Ecosystem proximity · Linked innovation

In 1909 on the outskirts of Jaffa, 66 families with an entrepreneurial spirit and great dreams founded the city of Tel Aviv. In 2009, 63 Israeli companies traded on the Nasdaq, more than the combined number of those from Europe, Japan, Korea, India, and China.

Tel Aviv is the second most populous city in Israel. With a vibrant and creative environment, it attracts successful young entrepreneurs. This, together with government initiatives, has turned this city into an ecosystem of innovation with world-class capabilities.

It currently has the world's highest start-up rate per capita. In addition, it is the technological capital of a country that invests a higher percentage of its GDP in R & D than any other.

Public programs, such as Yozma, have had a tremendous impact. This program was launched to support the creation of the country's venture capital market. In this program, the government contributes $100 million to finance high-tech start-ups. The money comes from 10 funds in the following way: (1) the government contributes up to 40% of each fund, which means that the private sector will disburse at least 60%, which is $150 million; (2) the 60% from the private sector must come not only from Israeli companies but also from foreign capital; (3) the funds have the option of buying the state's stake, giving them the choice of becoming 100% private funds; and (4) the government does not participate in fund management [1].

This ecosystem has dynamic agents such as Ramot, the "technology transfer arm" of Tel Aviv University. Ramot is responsible for starting, promoting, leading, and managing the transfer of new technologies from the university's research centers to the market and for protecting and marketing their discoveries.

There is also the Tech Incubator Program, an initiative launched to harness the knowledge of scientists who have immigrated from the Soviet Union. It invests in high-risk projects that would have difficulty finding private investors.

Then there is the Global Enterprise R & D Cooperation framework. This program promotes the collaboration of multinational companies in R & D projects with Israeli partners and focuses its efforts on areas within Israel.

However, are all innovation ecosystems as prepared as Tel Aviv? Unfortunately, they are not. So how can a research center monetize discoveries in an innovation ecosystem that is not so developed? What are the main challenges to tackle in interaction with the ecosystem?

6.1 From Causes of Broken Innovation to Best Practices for Linked Innovation

Do you lack innovation catalysts inside or outside your organization? This was a possible symptom identified in the case of broken innovation in the prioritization of catalysts during the three stages of the innovation funnel.

In the first case, you are probably prioritizing internal proximity, assuming that the whole innovation ecosystem is inside your organization. In the second case, you are probably prioritizing external proximity, assuming that the whole innovation ecosystem is near your organization.

In the analyzed cases, five causes were identified as the trigger of the symptoms of broken innovation in the innovation ecosystems (all stages). These five aspects are: lack of understanding of the innovation ecosystems, internal or external gaps, a lack of resources internally, and a low level of interaction in your ecosystem.

First, a lack of understanding or mapping of the innovation ecosystem happens when research centers have no clear idea or they have wrong assumptions about who the key stakeholders are within their innovation ecosystem. This affects not only quantity and dispersion but also quality.

For example, if I have to partner with a local research institution to apply for public funds, what are the most developed research centers in the specific field? What are the highest growth start-ups in my sector close to me geographically? Which companies are investing the most in R & D in my field of expertise surrounding my research center? And so on. The lack of information creates uncertainty when it comes to deciding among several partnerships, slowness in identifying opportunities and difficulty in prioritizing the 20% of stakeholders that are making 80% of the impact in the ecosystem.

According to the Global Entrepreneurship Monitor, 100 million businesses are created every year, meaning three every second [2]. Therefore, tracking such a volatile environment is not an easy endeavor.

A second factor is gaps in the internal innovation ecosystem. Getting initiatives approved is one thing. Getting initiatives implemented is another thing. Although we have seen the importance of decision-makers in these complex organizations (See Sect. 5.9), there is another problem in understanding and connecting with the other key stakeholders in research centers and in covering any gaps there may be in your institution.

A third cause is gaps in the innovation ecosystem (outside the research center), which means no proximity to certain actors. In this case, after getting an updated complete map, we identify the presence of spots or weaknesses in the innovation funnel, in terms of quantity or quality, or a lack of proximity to those agents. This could happen if, for instance, there is no support from government agencies to carry out the technology transfer process from research centers to the market, if there are not enough big corporations or budgets to invest locally in research projects, if the quality of the researchers is not so high or if the resources at universities are scarce, if the other agents of your innovation ecosystem are far from your research center, and so on.

The fourth case concerns a lack of resources in your innovation ecosystem (internally) and difficulty in retaining talent. Research centers advance in their roles with scarce resources: a limited budget, a reduced team of employees, and pressure from the leadership team to maximize output.

Fifth, there are few interactions among agents. In this scenario, although you have a complete innovation ecosystem, you do not interact with the other agents of your innovation ecosystem either internally—within your research center—or externally.

These causes were more or less common depending on the center type—its age and orientation. In the following sections, we will look at the nine kinds of best practice that solve each of the identified problems (See Table 6.1).

Table 6.1 All stages—innovation ecosystems: leveraging internal and external agents (overview)

Orientation and age of the center				Causes of broken innovation	Best practices for linked innovation
Research young	Research mature	Innovation young	Innovation mature		
x	x	x		(a) Agents: lack of understanding or mapping of the innovation ecosystem	(a1) Map the key stakeholders of your innovation ecosystem (a2) Identify the quality of the agents and the ecosystem
x		x		(b) Agents: gaps in the innovation ecosystem (internally)	(b1) Map and connect with your organizational chart, influencers and advisors
x	x	x	x	(c) Agents: gaps, little development or no proximity to the agents of the innovation ecosystem (externally)	(c1) Adapt your commercialization model to the characteristics of your innovation ecosystem (c2) Connect virtually with disperse agents

(continued)

Table 6.1 (continued)

Orientation and age of the center				Causes of broken innovation	Best practices for linked innovation
Research young	Research mature	Innovation young	Innovation mature		
x	x	x	x	(d) Agents: no resources within your innovation ecosystem (internally) and difficulty in retaining talent	(d1) Crowdsource, outsource or delegate the areas of your value chain that are not in the core of your business model and specialize each function (d2) Capitalize on aging
x		x		(e) Relationships: few interactions among agents	(e1) Move from academics to entrepreneurial academics (e2) Recognize academic entrepreneurs before they leave

Source Prepared by the author

6.2 Map the Stakeholders of Your Innovation Ecosystem

Who should I choose as a partner to apply for public funds? Who can complement my current weaknesses in technology transfers? Is there any gap in the innovation funnel?

The first step to answer all these questions is to draw a who's who diagram of the innovation ecosystem. In other words, which stakeholders are having the greatest impact, are more influential, and are driving the conversation?

In order to draw the diagram, you can start with the three big groups: university, industry, and government. A research group at Stanford University has called this model the Triple Helix, explained as the evolution from "a dominating industry-government dyad in the Industrial Society to a growing triadic relationship between university-industry-government in the Knowledge Society" [3]. This model can be applied at the regional, national, and international level of the innovation ecosystem (Fig. 6.1).

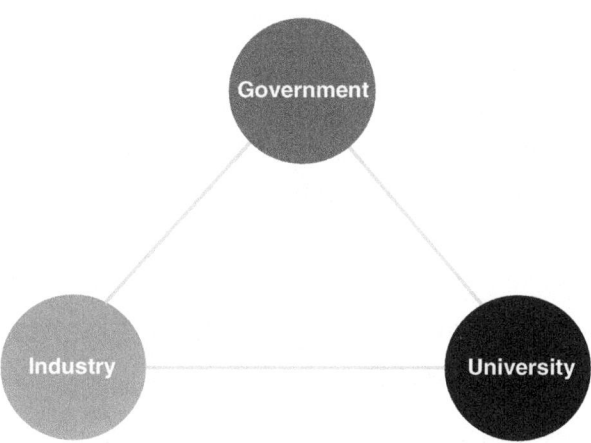

Fig. 6.1 Innovation ecosystem (agents and relationships): the triple helix model. *Source* Prepared by the author

The Triple Helix thesis is that the potential for innovation and economic development in a knowledge society lies more in the university and the interactions among university, industry, and government to generate new formats for the production, transfer, and application of knowledge, leveraging the creativity of each institutional sphere [4].

Following this model, we can map the innovation ecosystem of Tel Aviv in a network of agents and relationships (See Fig. 6.2). This exhibit gives us an overview of the interactions and the interconnections between actors.

This map shows us the quantification and the interconnection between the agents. The Israeli government facilitated the whole innovation process through the effective provision of infrastructure and regulations. It financed the Office of the Chief Scientist, the Global Enterprise R & D Cooperation program and tech incubators.

The university and research institutions undertake research and facilitate academic and economic growth through the creation, adaptation, and transfer of knowledge. These entities included Tel Aviv University, the national agency MATIMOP (Israeli Industry Center for R & D), the technology transfer arm Ramot, along with other research and development centers.

The companies that help to capture the value of the discoveries in the market are Israeli companies, foreign companies, and multinational corporations. Aside from the big corporations, there are more than 600 start-ups, in addition to incubators, accelerators, and several investment mechanisms such as business angels, venture capital funds, and micro-capital funds.

6.3 Identify the Quality of Your Agents and of Your Ecosystem

Do you know which agents are having the greatest impact? For example, do you know which business angels are achieving the highest rate of return on investment, which universities have the best rankings, which corporations are investing the most in R & D as a proportion of

6 All Stages: Innovation Ecosystem—Qualifying … 123

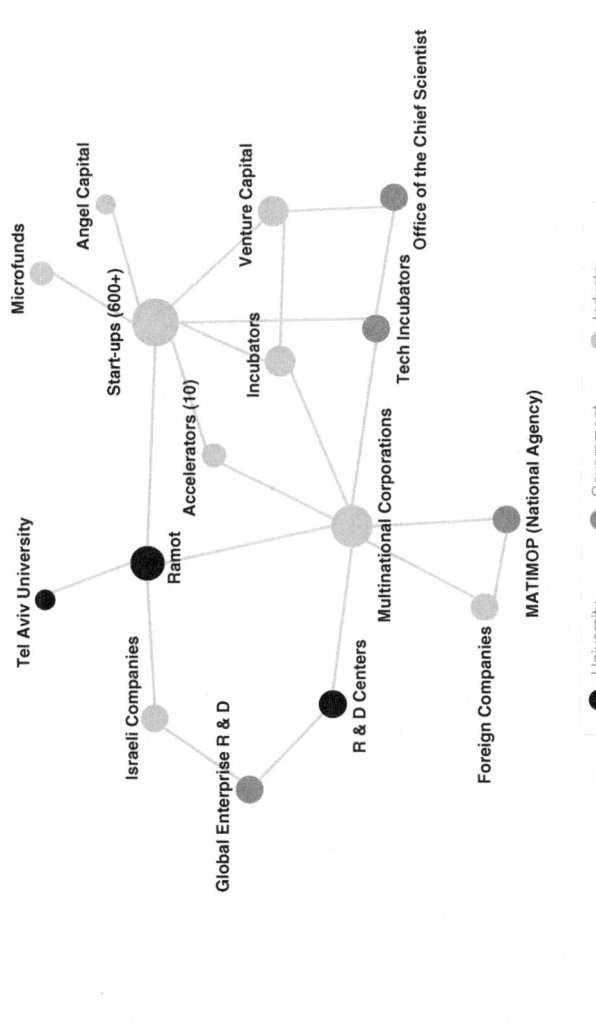

Fig. 6.2 Innovation ecosystem of Tel Aviv. *Source* Prepared by the author from several sources

revenue? Do you know the strengths and weaknesses of your ecosystem? If the answer is "no" then, after the first step of identification and quantification, the next step is qualification.

Not all countries have the same innovation ecosystem as Switzerland, Sweden, the United Kingdom, the United States, Finland, and Singapore. Those countries occupied the top six positions in the Global Innovation Index of 2016 and preceding years [5].

Such rankings can be used as a fast and easy way to understand which key stakeholders should be kept in mind when making decisions. Some databases are available online to qualify the agents of the ecosystem such as the Scimago Institutions Rankings [6] and Excellence Mapping [7].

By way of example, I illustrate the qualification of the research institution headquarters in the region of Catalonia, Spain (See Table 6.2). This table includes the type of institution (i.e., research based with a black circle, technology based with a white circle, transfer based with a grey circle), the qualification (according to several rankings—the better ranked the center, the longer the bar—and according to whether a research institution has received the Spanish Severo Ochoa grant).

Now when you want to choose a partner for a collaboration, leverage the ecosystem, or redirect a project to another stakeholder, you are able to identify the right place. Following this principle, Apple has created a pool of research centers in the country of the majority of its hardware manufacturing (China), reducing the gap between research and production [8].

After the identification and qualification of agents, the next step is a holistic view of the whole innovation ecosystem to understand the weaknesses and strengths. This overview will help us in decisions such as how much we leverage private compared with public funds, the size of our research teams, and our collaborative business model.

To measure the development of the innovation ecosystem, I suggest a simplification of the Global Innovation Index variables. This index shows the key variables of 128 innovation ecosystems, at a country level. It is updated annually by INSEAD, Cornell University, and the World Intellectual Property Organization. I have provided a simplified version for the purposes of this book (See Table 6.3).

Table 6.2 Quantification (on the map) and qualification (in the list) of research centers in the innovation ecosystem (Catalonia, Spain)

Type	Name	Qualification	S. Ochoa
○	Leitat Technological Center		
●	Barcelona Supercomputing Center		▪▫▫
◐	Technological Center of Nutrition and Health		
●	Center for Genomic Regulation		▪▫▫
●	Esadecreapolis		
●	Barcelona GSE Center		▪▫▫
◐	Fostering Arts and Design		
○	Barcelona Media		
○	Cetemmsa Foundation		
○	Ascamm Technology Center		
○	Barcelona Digital Center		
◐	i2Cat Catalonia Digital Lab		
●	Institute of Chemical Research of Catalonia		▪▫▫
●	Catalan Institute of Nanoscience and Nanotechnology		▪▫▫
●	The Institute of Photonic Sciences		▪▫▫
●	High Energy Physics Institute		▪▫▫
●	UB Institute of Biomedical Research		▪▫▫
●	August Pi i Sunyer Biomedical Research Institute		
●	Science and Technological Park Orbitral 40		
●	UB Institute for Research in Biomedicine		
●	Lleida Agri-food Science and Technology Park		
●	Science and Technological Park of the University of Girona		
●	Barcelona Biomedical Research Park		
●	UAB Research Park		
●	UPF Research Park		
●	UPC Barcelonatech Park		
●	Technova Barcelona		

Source Prepared by the author using data from Prats, J., Siota, J., and Gironza, A. *2033: compitiendo en innovación*, 38–39 (IESE and PwC 2015) [1]

This Global Innovation Index study includes updated data points for the five pillars of every innovation ecosystem: institutions (i.e., political environment, regulatory environment, business environment), human capital and research (i.e., pretertiary education, tertiary education, research and development), infrastructure (i.e., digitization, general infrastructure, ecological sustainability), market sophistication (i.e., credit, investment, trade), and business sophistication (i.e., knowledge workers, innovation linkages, knowledge absorption).

Table 6.3 Simplified visualization of the Global Innovation Index (South Africa)

Variables	Score (0–100)	Rank (0–128)
1. Institution	**69**	**46**
Political environment	55	56
Regulatory environment	75	38
Business environment	78	38
2. Human capital and research	**33**	**55**
Education (before tertiary level)	44	74
Tertiary education	27	89
R & D	28	40
3. Infrastructure	**37**	**85**
ICT	40	84
General infrastructure	39	54
Ecological sustainability	34	101
4. Market sophistication	**59**	**17**
Credit	40	44
Investment	66	10
Trade, competition, and market scale	70	33
5. Business sophistication	**32**	**56**
Knowledge workers	36	70
Innovation linkages	32	59
Knowledge absorption	29	63
6. Knowledge and technology outputs	**25**	**63**
Knowledge creation	16	52
Knowledge impact	35	67
Knowledge diffusion	23	73
7. Creative outputs	**27**	**77**
Intangible assets	38	83
Creative goods and services	25	55
Online creativity	5	77

Source Prepared by the author using data from The Global Innovation Index 2016: Winning with Global Innovation (Cornell, INSEAD, and WIPO 2016) [5].
Note The column scores have been rounded up or down

If you are managing a research center in South Africa, for example, and you want to spot quickly the center's main strengths and weaknesses, by looking at the study you will discover a country positioned 54 out of 128 with a medium developed institutional environment (ranked 46 out of 128) with political instability and low-government effectiveness. However, it is easy to pay taxes there, for which it is ranked 19. In terms of human capital and research, expenditure on education before the tertiary level is 6.1% of GDP but there is a low level of tertiary enrolment (19.7%).

Infrastructure is below average (ranked 85 out of 128), with a low level of e-participation and low-ecological sustainability. South Africa is well positioned for market sophistication, being ranked 17, with good development of domestic credit to the private sector compared with 0% of microfinance loans. There are good levels in the investment subpillar.

Interesting business sophistication gives it a ranking of 56, including average knowledge absorption levels but with a high level of intellectual property payments (1.5% of total trade).

In terms of output, although the country has very low levels of ICT service exports, the rate of new businesses is quite high (ranked 18 out of 128). Finally, the country is poorly positioned (77) for creative output.

Now that you have mapped and understood your innovation ecosystem, you will find three possible configurations: driven by government, driven by industry, and driven by university-industry-government.

In the first case, the government drives universities and industry, limiting their innovation processes (e.g., Russia). In the second case, industry is the driving force, the university is a supplier of skilled human capital, and government is a regulator (e.g., the United States). In the third case, all three agents join forces with an aligned goal, favoring an environment for innovation (e.g., Sweden) [9].

6.4 Map Your Center and Connect with Decision-Makers, Influencers, and Advisers

Are you tired of launching initiatives that never get implemented? Is it becoming the norm for initiatives to take twice as long as you planned to be executed? Do they even sometimes get lost in long decision-making processes through several business units?

According to an article by Gilad Raichshtain on the Salesforce website, an average of 84 days is needed to convert 13% of leads into opportunities in business-to-business (B2B) sales. Then the decision process is much faster: it takes 18 days to convert an opportunity into a deal, but only 6% of opportunities end being converted to deals. Although some directors know these numbers, they may not realize that selling internally, especially in bureaucratic and political institutions, is a similar challenge to a B2B sale, except that the customer is your own institution [10].

In this complex environment, three tools help research center managing directors to accelerate the execution of internal processes needed for the commercialization of discoveries.

In this complex environment, three tools help research center managing directors to accelerate the execution of internal processes needed for the commercialization of discoveries, as we have seen in the case of the Merck's Word Wide Licensing and Knowledge Management group (See Sect. 5.9).

The first is the organizational chart, containing not only those in official positions but also those who are really driving your institution albeit unofficially (e.g., thought leaders, decision-makers, and influencers). Thought leaders are those who may help you understand aspects that are outside your area of expertise, decision-makers are those who will approve or reject your proposals, and influencers may help you accelerate the process or convince decision-makers.

The identification of unwritten rules was recognized as important for this aspect—these rules that are unexpressed but usually applied in the organization constitute a kind of internal politics.

I give an example of a research center's organizational chart (See Fig. 6.3). The chart includes rows showing who reports to who. This chart is a simplified adaptation of a common organizational chart used in research centers [11].

The second step is to identify the KPIs of each stakeholder, in this case not only of the decision-makers (See Sect. 5.9) but also of the other agents on your map. In the example, the KPIs are written below the role description with a number in brackets.

The final step is to network with each of them and maintain a good relationship.

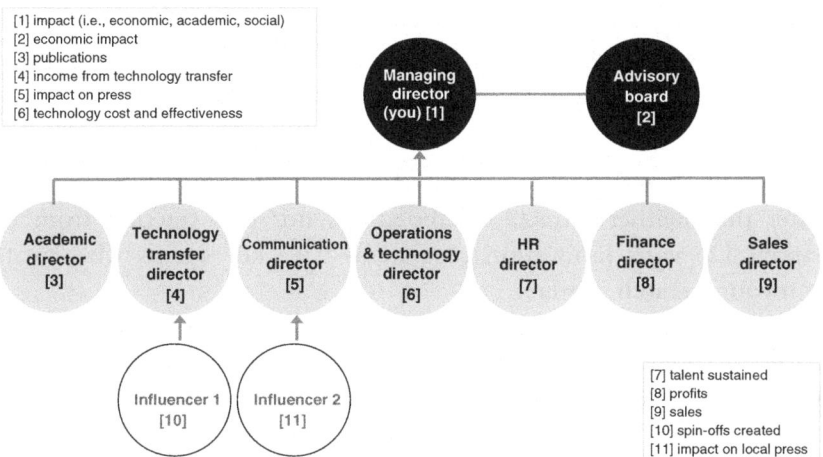

Fig. 6.3 Example of a research center's organizational chart and KPIs. *Source* Prepared by the author

A diverse network in the crucial business area, not only at the leadership level, helps those in the research center leadership roles being analyzed to understand the internal processes and speed up the execution of initiatives.

In B2B sales processes, many methodologies can be used to conceptualize and articulate these kinds of organizational charts such as the Conceptual Selling method of Miller and Heiman. These methods may help you move in those complex B2B sales within your own institution.

In this case, the Miller and Heiman method suggests that sales professionals (i.e., you) should focus on the basics of the client's buying process (i.e., that of your research center) and manage the multiple decision-makers involved in the process, rather than going straight in with their sales pitch. This methodology requires a great deal of listening to gather information and it also requires the client (i.e., the internal decision-maker) to be provided with relevant information and a multilevel commitment to be attained [12].

6.5 Adapt Your Commercialization Model to the Characteristics of Your Innovation Ecosystem

Are you implementing a commercialization model at your research center that neither attracts collaborations nor gets traction from the market? Do your innovation ecosystems have key agents who are far from your research center?

One night at Boston, I was having a quick dinner in the cafeteria of the Ritz Carlton Hotel with seven colleagues coming from several universities. One of them, who is in the leadership team of the MIT Regional Entrepreneurship Acceleration Program (REAP), explained how the program works.

Since MIT already has a well-developed ecosystem, the university started this initiative to help regions outside Boston accelerate economic growth and promote social progress through innovation-driven entrepreneurship. Partner regions form multidisciplinary teams and commit to a 2-year learning engagement in which teams work with MIT faculty and the broader MIT REAP community to build and implement a customized regional strategy to enhance their ecosystems [13].

Nevertheless, different innovation ecosystems do not have the same agents and characteristics. For instance, while gross expenditure on research and development (GERD) is decreasing in the United States, China's GERD is increasing [14]. As a consequence, what works in the United States will not necessarily work in China. In other words, best practices should be not adopted but adapted to your own characteristics.

This adaptation of your commercialization model and the best practices that others are implementing are especially important when some innovation ecosystem's agents are nonexisting, less developed, or far away. This adaptation includes aspects such as the right size of research teams so that companies can absorb the generated knowledge (See Sect. 5.8), the appropriate investors to fund the spin-offs launched from your center.

In terms of proximity, research centers sometimes, but not always, need all the catalysts of the innovation ecosystem near them. Often they are able to operate in other geographic areas and can fill any gaps just by adapting their business model to choose the best catalysts.

For instance, business angels in the regions of Spain are regarded as investing almost exclusively in projects in their own region. As a consequence, being positioned geographically close to business angel networks may increase the chance of being funded.

However, the industry does not need to be geographically close during the commercialization stage. The proximity affects only the type of collaborations made due to geographic location. For example, research centers near the market may launch spin-off models, leveraging the absorption of knowledge by commercializing discoveries in the nearby market, while research centers that are not near the market may apply a licensing model that can be applied more easily to a faraway market [1].

Research centers apply several models depending on their taxonomy. For instance, 25 companies are spun off from MIT each year, on average. Also, technology originating from research at Stanford University has fueled the growth of many companies in Silicon Valley.

Examples are not restricted to the United States—Sweden's Chalmers University produces 10–15 spin-offs each year, many of them small consulting and computer companies, contributing $100 million to the local economy each year [15]. In the United Kingdom, "many of the 450 high-technology companies in its region derive from Cambridge University spin-offs, which provide more than half the manufacturing jobs in the region [16]."

6.6 Connect Virtually with Dispersed Agents

Do you lack other experts in your field in your own geographic area? Do you have gaps in your innovation ecosystem?

Digitization has made life easier for researchers. There are new ways to work together in parallel on the same project using different computers (e.g., with the software Authorea) and to optimize the management of a project's literature review and citation process among research groups (e.g., with the software Mendeley, acquired for more than $70 million by the educational publisher Elsevier) [17]. Digitization has

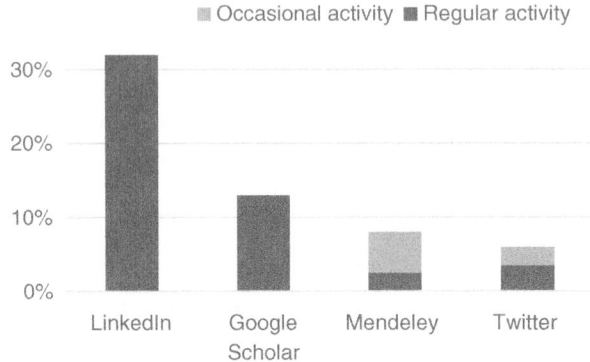

Fig. 6.4 Use of digital tools by researchers. *Source* Prepared by the author using data from Tinkler et al. *The impact of the social sciences: how academics and their research make a difference* (SAGE Publications, 2014) [18]

also made it easier to create international think tanks and communities of experts in several fields (Fig. 6.4).

Digital technology is set to radically modify the way science is conducted and the way research results are disseminated. Open science—as a new paradigm—is emerging. This trend encompasses open access to scientific journals, open research data, and open collaboration enabled by ICT. This field of research e-infrastructure is probably one of the areas that has benefited the most from increased international policy coordination in recent years [14].

Several experts have highlighted the benefits of opening up the innovation process because, by considering the solution knowledge of a large number of individuals, new creativity can be brought into the organization [19, 20].

There are several examples of successful research networks, such as the European Forum for Entrepreneurship Research (EFER). The forum aims to put entrepreneurship on the road map of European higher education and develops a network of first-class European professors and active entrepreneurs [21].

Canada's Networks of Centres of Excellence (NCE) has brought together widely geographically dispersed academics and firm research units, motivated to work together. By harnessing the best talent in the natural sciences, engineering, social sciences and health sciences, NCE programs help build a more advanced, healthy, competitive, and prosperous country [22].

A final example is the Rolls-Royce Research and University Technology Centres, an established global network of research centers. Each center addresses a key technology. They collectively tackle a wide range of engineering disciplines [23].

In summary, now more than ever, it is easier to achieve the virtual congregation of geographically dispersed groups of experts from university and industry around common research themes, with government support. So you can leverage those virtual connections to cover gaps in your ecosystem.

6.7 Crowdsource, Outsource, or Delegate Areas of Your Value Chain that Are not Core to Your Business Model

Have you received approval for a lower budget than you expected for this course? Do you need more human resources to execute the whole range of initiatives? Have you identified in your strategic plan for this course that your team will have to work 1.5 times the average number of hours per day? Are the quarterly economic reports not ready to be delivered?

Global R & D capacity has doubled in the past 15 years, with a growing share of business expenditure compared with public and private funding. However, it has become more difficult to reach those resources in several segments.

Corporations with revenues greater than $100 million reduced their R & D intensity (R & D spend divided by revenue) by 5.6% on average, whereas capital intensity in those companies fell by only 4.8% and advertising intensity actually increased by 3.4% [24].

However, public funds are under pressure, and weak economic conditions have reduced tax revenues and public budgets. Additionally, the percentage of public R & D funding in some countries has decreased compared with worldwide public R & D funding. This is the case in the United States, Japan, and the European Union member states [14].

Scarcity of resources is a common theme among research centers: tight budgets, reduced teams and the expectation of many outputs. What is the solution?

At the end of the day, you have defined your collaborative business model, which relies on a core business and competitive advantage. However, you can crowdsource, outsource, or delegate to less expensive profiles the tasks that are not in those core boundaries [25].

In professional service firms, on average one senior member requires two middle-ranking members and six junior members. This ratio of 1:2:6 enhances the leverage of the structure [25].

During your research projects, research leaders probably do not have to spend time gathering data but framing the overall research question and analysis. You can delegate those tasks to research assistants or intern students. This is the common talent pyramid structure applied in the Henderson Institute at The Boston Consulting Group, the Institute for High Performance at Accenture, and the Health Research Institute at PwC.

Another example is to leverage fourth- and fifth-year Ph.D. students for literature reviews. Giving them tools to optimize the process and providing them with the framework to go into details more quickly may save research teams a lot of time.

In the literature review, Ph.D. students could be assessed on keywords to use in the initial search, which journals to prioritize, etc. In terms of the use of software, they could optimize the literature track and management process, such as with Mendeley. The students could be given specific tips for the search (e.g., using words such as "state of the art" to find faster peer-reviewed articles that have already covered the same topic).

A final example is to apply the search model (See Sect. 5.2.9) to identify potential leads to be given ad hoc research projects from industry.

6.8 Capitalize on Aging

Are the experts you want to hire too expensive for your annual budget? Do you fail to find the right experts at the right time? Would you like to have experienced coaches for your spin-offs?

When I asked other professionals, "What do you think is the highest-growth age range in which founders create new ventures in Silicon Valley?" I typically got the answer of between 20 and 30. The respondents thought that entrepreneurs were enthusiastic young professionals or university dropouts. However, this is far from reality.

According to the Index of Growth Entrepreneurship—released annually by the Kauffman Foundation and focusing on education and entrepreneurship—the highest growth of entrepreneurial activity in Silicon Valley takes place among 55- to 64-year-olds [26].

Several centers were leveraging professionals in this age range. These were experienced professionals who had succeeded in the past and were nearing the end of their professional careers. They were searching for more stable engagements and had extensive networks and experience [27].

One success story is the WeGrow program, launched by IESE Business School in Catalonia. The initiative connects alumni: high-growth entrepreneurs and experienced leaders with a track record. During the 10-month program, each mentee meets on-site with a group of three mentors to discuss a management challenge for 1 hour each month [28].

6.9 Move from Researchers to Academic Entrepreneurs

Are you experiencing a low degree of interaction between your scientists and industry? Is it difficult to coach your researchers or get them to understand the economic metrics?

The Entrepreneurial University is a central concept in the Triple Helix model of university-industry-government (See Sect. 6.4.). It takes a proactive attitude in not only creating new knowledge but also putting knowledge to use. It follows an interactive rather than a sequential approach. In this model, the industry participates in upper levels of training and knowledge sharing; the government acts not only as a regulator but also as a venture capitalist and public entrepreneur, and the university fosters collaborations to combine and capture value from knowledge.

As a consequence, we have the academic "third mission" of being involved in socioeconomic development in addition to the traditional missions of teaching and research (See Sect. 3.2), creating collaborative links with the other innovation actors, especially in advanced areas of science and technology. Universities increasingly become the source of regional economic development and academic institutions are reoriented for this purpose [9, 29, 30].

Two examples of the application of this concept come from San Diego and Stockholm.

The San Diego branch of the University of California, created in 1950, became home to a leading high-tech complex and contributed to the transformation of San Diego from a naval base and military retirement community to a knowledge-based conurbation.

The coalition of academic, business, and political leaders attracted leading researchers in fields with commercial potential—such as molecular biology—as a strategy for economic development. The strategy of the San Diego campus was replicated by the Merced campus, recently established as an "entrepreneurial university." This institution promotes high-tech development in an agricultural region and creates knowledge assets, attracting new investment and creating new value.

In the second example, a network of existing knowledge-based organizations created a coalition to become internationally competitive. The Stockholm School of Entrepreneurship was created as a joint initiative of Stockholm University, the Royal Institute of Technology (KTH), and the, recently included, Royal Art College.

So how do you select and transform your current scientists into entrepreneurial scientists?

The entrepreneurial scientist concept combines academic and business elements. It simultaneously addresses advancing the frontiers of knowledge and mining its practical and commercial results for industrial and financial returns. The foundation is the polyvalent nature of knowledge, which is at the same time theoretical and practical, publishable, and patentable [31].

Achieving a team of entrepreneurial scientists means placing importance on the selection process as well as understanding the preferences and interests of the scientists.

There are several academic entrepreneurial styles. The first style is having a direct interest in forming spin-offs and taking a leading role in this process. The second is handing over these results to a technology transfer office. The third is playing a supporting role as a member of a scientific advisory board. And the fourth is having no interest in entrepreneurship but viewing firm creation as a useful source for developing needed technology [31].

Groups of complementary entrepreneurial individuals are particularly visible in high-tech. A high-tech start-up typically takes off with the support of individuals with technical and business expertise backed by an experienced entrepreneur, together constituting a kind of collective entrepreneur, because rarely does an individual embody all of these required elements.

Nevertheless, these collaborations are influenced by cultural aspects. For instance, while in the United States there is a strong ideology of individual entrepreneurship that commonly suppresses the contributions of collaborators, in Nordic countries there is a strong collaborative mind-set, keeping individuals from starting entrepreneurial ventures unless they are backed by a group [32].

In summary, a clear identification of the preferences of the scientists and a deep understanding of their cultural mind-sets will contribute to their transformation into entrepreneurial scientists.

6.10 Recognize Academic Entrepreneurs Before They Leave

Have you identified entrepreneurial scientists in your research center and yet don't know how to keep them? Are your science entrepreneurs sometimes drowning in the deep level of corporate politics, the high pressure of short-term results, and tight budgets?

Beyond the gifts of nature, we are surrounded by creations that came from science. Nevertheless, scientists are rarely the ones who make money from the advances that shape our world. While Silicon Valley is churning out billionaires under 30 (just check the latest "Forbes

30 under 30" or "Fortune 40 under 40"), no Nobel Laureate has ever become a billionaire [33].

Additionally, any managing director at a research center can tell you that coming up with ideas is not the problem. The real issue is selecting, testing, and executing the best ones. To build this engine, research centers need the skills to transform ideas from science entrepreneurs who are willing to take a step forward in internal politics.

You may already have a few science entrepreneurs in your research center, but most of them are hidden. When you find them and support them properly, magic can happen.

Science entrepreneurs can improve your research center more quickly and effectively than others because they are self-motivated free thinkers and masters at navigating around bureaucratic and political inertia.

In organizations, for every 200 employees there are at least 10 natural science entrepreneurs. One out of those 10 is a great science entrepreneur who can build the next business for your research center or for the market [34].

Some 70% of successful science entrepreneurs got their business idea while working for a previous employer. These talented individuals left because the environment did not have an intrapreneurial process to pitch their ideas or their bosses did not support them in the new ventures. Smart scientists have left research centers to start their own ventures because their institutions did not believe in intrapreneurship as a critical tool for growth [35].

According to Harvard Business School Professor Vijay Govindarajan, there are six DNA patterns for identifying these successful intrapreneurs. First, money is not their core measurement. They are searching for the freedom to create. Rewards are simply the scorecard of how well they are doing it.

Second, they use continuous strategic scanning. They are always thinking about how to leverage the current situation and which step to take next. Third, they are like greenhouses; they cultivate their ideas privately, letting them take shape before showing them to the world. Fourth, they are visual thinkers, using a combination of brainstorming, mind mapping, and design thinking [34].

Fifth, they know how to pivot their ideas, and how to resuscitate a dead one. One example would be what Jeff Bezos did with Amazon, changing its business model from being the world's largest online market, selling everyone else's products, to selling its own hardware, the Kindle, and achieving more than 60% of the e-reader market. And, sixth, they are authentic and well-rounded people—great leaders who combine confidence with humility.

These aspects are very much aligned with the conclusions of a recent study explaining the motivations of academic entrepreneurs (See Fig. 6.5).

However, once you identify the golden stone, there is a high risk of losing it. How do you keep them?

First, prepare an entrepreneurial environment in your research center that allows for the development of new ideas and technologies and a reframing of the company mind-set. This environment has several characteristics. You may start by creating a sense of ownership, allowing employees to make decisions that have an impact on their work and leading them to view the research center as their own.

Second, be flexible. Do not rely only on proven methods and be prepared to accept new systems to solve problems. Third, accept faults.

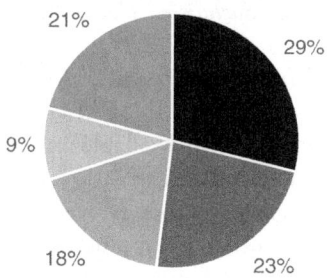

Fig. 6.5 Motivational measures of academic entrepreneurs. *Source* Prepared by the author with data from Hayter [36]

Otherwise, it will be impossible to innovate. Find a way to reduce to an acceptable level the researchers' personal risk when they are innovating. If not, they will be restrained by the risk of failure.

Fourth, give them freedom. In general, their primary motivation is to influence—through development and diffusion—with freedom, so enabling internal support from senior management so they can implement their ideas would help in this endeavor. Finally, reinforce them with incentives.

Some 1,400 companies and 40,000 jobs have resulted from the University of Cambridge's Science Park, the largest center for commercial research in Europe. The institution is just one in a wide pool of entrepreneurial universities [37].

In summary, to sustain your growth engine through innovation, you should detect and keep intrapreneurs in your organization by preparing the suitable environment for them.

References

1. Prats, J., Siota, J. & Gironza, A. *2033: compitiendo en innovación*. (PriceWaterhouseCoopers; IESE Business School, 2016).
2. Global Entrepreneurship Monitor. *2015/2016 Global Report* (2016).
3. The Triple Helix Research Group. Stanford University. *The Triple Helix Concept*. Available at: https://triplehelix.stanford.edu/3helix_concept. (Accessed 26 Feb 2017).
4. Schumpeter, J. A. *Capitalism, Socialism, Democracy* (George Allen and Unwin, 1942).
5. Dutta, S. et al. *The Global Innovation Index 2016: Winning with Global Innovation*. (Johnson Cornell University, INSEAD, WIPO, 2016).
6. Scimago. *Scimago Institutions Rankings*. Available at: http://www.scimagoir.com/rankings.php (2017). (Accessed 26 Feb 2017).
7. Bornmann, L., Mutz, R., Stefaner, M. & Moya, F. *Excellence Mapping*. Available at: http://www.excellencemapping.net/#/view/edition/2014/measure/top10/calculation/a_ohne_kovariable/field/materials-science/significant/false (2016). (Accessed 26 Feb 2017).
8. Gartenberg, C. Apple is opening two more R & D centers in China. *The Verge* (2017).

9. Etzkowitz, H. & Leydesdorff, L. The dynamics of innovation: from national systems and "Mode 2" to a Triple Helix of university-industry-government relations. *Science and Technology* **29**, 109–123 (2000).
10. Raichshtain, G. *B2B Sales Benchmark Research Finds Some Pipeline Surprises* (Salesforce, 2014).
11. Center for Research in Agricultural Genomics. *Organizational Chart* (2017).
12. Miller, R. B., Heiman, S. E. & Tuleja, T. *The New Conceptual Selling: The Most Effective and Proven Method For Face-To-Face Sales Planning* (Warner Business Books, 2005).
13. MIT University. *MIT Regional Entrepreneurship Acceleration Program (REAP)—MIT Innovation Initiative*. Available at: https://innovation.mit.edu/resource/mit-reap/. (Accessed 26 Feb 2017).
14. OECD. *Science, Technology and Innovation Outlook 2016* (2016).
15. McQueen, D. H. & Wallmark, J. T. University technical innovation: spin-offs and patents in Goteborg, Sweden, in *University Spin-off Companies* 103–115 (Rowman and Littlefield Publishers, 1991).
16. Roberts, E. B. & Malonet, D. E. Policies and structures for spinning off new companies from research and development organizations. *R & D Management* **26**, 17–48 (1996).
17. Lunden, I. *Confirmed: Elsevier Has Bought Mendeley For $69M–$100M To Expand Its Open, Social Education Data Efforts | TechCrunch. Tech Crunch (2013)*. Available at: https://techcrunch.com/2013/04/08/confirmed-elsevier-has-bought-mendeley-for-69m-100m-to-expand-open-social-education-data-efforts/. (Accessed 26 Feb 2017).
18. Tinkler, J., Dunleavy, P. & Bastow, S. *The Impact of the Social Sciences: How Academics and Their Research Make A Difference* (SAGE Publications, 2014).
19. Ye, J. & Kankanhalli, A. Exploring innovation through open networks: A review and initial research questions. *IIMB Management Review* **25**, 69–82 (2013).
20. Enkel, E., Gassmann, O. & Chesbrough, H. Open R & D and open innovation: Exploring the phenomenon. *R & D Management* **39**, 311–316 (2009).
21. European Forum for Entrepreneurship Research. *Who we are and what makes us tick*. Available at: http://www.efer.eu/about/. (Accessed 26 Feb 2017).
22. Government of Canada. *Networks of centres of excellence programs.*

23. Rolls-Royce. *Research and University Technology Centres*. Available at: https://www.rolls-royce.com/about/our-technology/research/research-and-university-technology-centres.aspx#/our-technology. (Accessed 7 March 2017).
24. Knott, M. The Trillion-Dollar R & D Fix. *Harvard Business Review* **90**, (2012).
25. Maister, D. H. *Managing the Professional Service Firm* (Free Press, 1993).
26. Morelix, A., Reedy, E. J. & Russell, J. *Kauffman Index—Growth Entrepreneursip* (Kauffman Foundation, 2016).
27. Bozeman, B., Fay, D. & Slade, C. P. Research collaboration in universities and academic entrepreneurship: The-state-of-the-art. *The Journal of Technology Transfer* **38**, (2013).
28. IESE Business School. *Business Leaders Mentor Startup Entrepreneurs*. Available at: http://www.iese.edu/en/about-iese/news-media/news/2016/april/business-leaders-mentor-startup-entrepreneurs/ (2016). (Accessed 26 Feb 2017).
29. Etzkowitz, H. *MIT and the Rise of Entrepreneurial Science* (Routledge, 2002).
30. Guerrero, M., Cunningham, J. A. & Urbano, D. Economic impact of entrepreneurial universities' activities: An exploratory study of the United Kingdom. *Research Policy* **44**, (2015).
31. Ranga, M. & Etzkowitz, H. Triple Helix systems: an analytical framework for innovation policy and practice in the knowledge society. *Industry and Higher Education* **27**, 237–262 (2013).
32. Freiberger, P. & Swaine, M. *Fire in the Valley: The Making of the Personal Computer* (McGraw-Hill, 2000).
33. Rathi, A. Why scientists make bad entrepreneurs and how to change that. *Quartz* (2015).
34. Govindarajan, V. & Desai, J. *Recognize Intrapreneurs Before They Leave* (Harvard Business Review, 2013).
35. Chamorro-Premuzic, T. *How Bad Leadership Spurs Entrepreneurship* (Harvard Business Review, 2012).
36. Hayter, C. S. In search of the profit-maximizing actor: motivations and definitions of success from nascent academic entrepreneurs. *The Journal of Technology Transfer* **36**, 340–352 (2011).
37. Davies, L. Hi-tech cluster keeps business booming in Cambridge. *The Guardian* (2012).

7

Conclusions

Abstract Finally, synthesizing the symptoms and causes of broken innovation, this chapter covers the four movements needed to advance from broken to linked innovation to become an economically sustainable research center while preserving research quality. First, choosing performance metrics for selecting research initiatives, changing your mindset from highlighting only academic or economic metrics to picking holistic metrics with academic, economic, and social impact. Second, understanding market needs, from following desires to applying design thinking. Third, collaborating with industry, moving from relationships for simply gathering data or selling research to collaborative business models with industry, universities, and government. And fourth, leveraging the center's innovation ecosystem, shifting from agent-proximity to agent-meritocracy prioritization.

Keywords Performance metrics · Market understanding · Design thinking · Collaborative business model · Innovation ecosystem · Economic sustainability · Research quality · Broken innovation · Linked innovation · Research center

Why do research centers so often fail to commercialize discoveries? The beginning of this book introduces the core challenge faced by research center managing directors: how to achieve economic sustainability while preserving academic quality.

Promising to drive economic sustainability in research centers while preserving research quality, the concept of linked innovation suggests a new path of conciliation between academic and economic orientations.

The symptoms for identifying broken innovation and the principals involved in moving to linked innovation provide a valuable compass at research centers in universities, industry, and government (See Table 7.1).

This book covers 28 mechanisms (See Table 8.2) and 12 business models to drive growth in research centers through the examination of performance metrics, design thinking, business modeling, and innovation ecosystems. These mechanisms are followed by successful research centers at universities (e.g., Harvard, Oxford, and Tel Aviv), in industry (e.g., Roche, Apple and JP Morgan), and in government (e.g., NASA, Fraunhofer, and the Chinese Academy of Science).

The results and recommendations are worthwhile not only for research centers but also for the entire range of agents in the innovation ecosystem surrounding the centers, including governments (policymakers), universities (contiguous business units), and industry (corporations, professional service firms, investors, etc.).

I hope this book will inspire executives and academics to create a knowledge-sharing network of research centers worldwide applying these principles, as well as further research with empirical data and business cases on additional mechanisms for implementing linked innovation effectively.

I would be glad to hear your feedback. Let's connect via LinkedIn and continue the conversation.

7 Conclusions

Table 7.1 From broken to linked innovation

	Stage 1: research	Stage 2: transformation	Stage 3: commercialization
Broken innovation	**Performance metrics** Economic vs. academic Are you facing a decline in research quality or in economic profitability?	**Market understanding** Assuming vs. following Are you coming up with products that no one wants to buy or that are outdated?	**Industry collaboration** Research vs. furtive Are you experiencing increased difficulty in monetizing your discoveries or in getting access to industry data?
		Innovation ecosystem Internal vs. external *Do you lack an innovation catalyst inside or outside your organization?*	
Linked innovation	**Performance metrics** Holistic impact Selecting performance metrics based on academic, economic and social impact	**Market understanding** Design thinking Translating discoveries into impact for the market through design thinking	**Industry collaboration** Collaborative business model Designing collaborative business models for university-industry-government relations
		Innovation ecosystem Meritocratic innovation ecosystem *Qualifying and leveraging the internal and external agents based on merit*	

Source Prepared by the author

8
Appendix

8.1 Motivations to Commercialize Discoveries

See Table 8.1.

8.2 Causes of Broken Innovation, Best Practices in Linked Innovation, and Case Examples

See Table 8.2.

Table 8.1 Motivations to commercialize discoveries

For universities	For industry
Teaching opportunities	Sourcing latest knowledge and technological advances
Funding and financial resources	Use of facilities and laboratory
Source of knowledge and empirical data	Personnel resources and cost savings
Enhancement with public funds from government	Risk sharing for basic research
Enhancement of reputation	Establishing long-term research projects
Job offers for researchers and students	Recruiting channel

Source Prepared by the author based on the analysis of the sample of research centers, the interviews and the literature review [1–11]

Table 8.2 Examples of each mechanism of linked innovation

Causes	Orientation and age				Best practices	Some case examples
	RY	RM	IY	IM		
Stage 1—research: the performance metrics						
(a) Project: nonholistic prioritization	x	x	x		(a1) Prioritize research projects based on holistic impact	MIT Deshpande Center National Aeronautics and Space Administration General Electric McKinsey Global Institute
(b) Interconnectivity: lack of knowledge sharing		x		x	(b1) Research map: map the research interests of each researcher and the research projects of the center	Harvard Business School's entrepreneurship unit University of Cambridge's Department of Engineering Hoffmann-La Roche World Bank Google Health Canada
(c) Leadership: lack of nonacademic experience	x	x	x		(c1 and d1) Use professional recruitment for academic and executive directors	Barcelona Super Computing Cornell Tech University of Melbourne's School of Chemistry
(d) Rigor: lack of academic experience	x		x	x	(c2 and d2) Attract and recruit an international advisory board	Pittsburgh High-Tech Council Petropolis Technopole Board of the Recife Brazil Science Park

(continued)

Table 8.2 (continued)

Causes	Orientation and age				Best practices	Some case examples
	RY	RM	IY	IM		
Stage 2—Transformation: market understanding						
(e) Information: unknown market needs	x	x	x		(e1) Use design thinking and the market map: translate and map consumer needs	IDEO United Kingdom's Design Council Stanford University's Institute of Design Industrial and Commercial Bank of China's Research Center Auckland Savings Bank's Innovation Lab J.P. Morgan Chase & Co.'s Technology Hub in the United States, BNP Paribas' Innovation Center Wells Fargo's Research Group Royal Bank of Canada's Innovation Center
					(e2) Follow lean research principals: maximize your learning speed and minimize your testing cost	MIT Design Lab Tufts Fletcher School Feinstein International Center Root Capital
(f) Context: unknown business arena	x	x	x		(f1) Complement the current services of your knowledge and technology transfer office	Humboldt Innovation Harvard Innovation Lab IESE Business School SAP Next-Gen Wharton's Mack Institute for Innovation Management Buckman Laboratories
(g) Diversity: lack of academic or executive profiles	x	x			(g1) Create diversified teams of executives (with MBAs) and academics (with Ph.Ds.)	Center for Genomic Regulation IBM's nanotechnology center Deutsche Telekom Laboratories
(h) Mentoring: uncoachable researchers			x	x	(h1) Include the indicator "coachable" in the recruitment, evaluation and incentive scheme of academics	Johns Hopkins University's economic development office MIT Deshpande Center for Technological Innovation

(continued)

Table 8.2 (continued)

Causes	Orientation and age				Best practices	Some case examples
	RY	RM	IY	IM		
Stage 3—Commercialization: industry collaboration						
(i) Business modeling: not clear or undefined	x	x	x		(i1) Understand and design a clear collaborative business model (i2) Select the collaborative business model that fits the orientation and age of your center	National Science Foundation of Engineering Research Centers Hewlett-Packard's Laboratories Nokia and UC Berkeley's Department of Traffic Engineering Deutsche Bank's research group The Economist Group Jigsaw (the former Google Ideas) Samsung Economic Research Institute PwC—Google McKinsey's Growth in Tech practice SAP Intel Chinese Academy of Sciences' Institute of Biophysics Ford Adobe Fraunhofer Institute for Solar Energy Systems ISE—CGA Technologies Lawrence Berkeley National Laboratory Oxford University Innovation (former Isis Innovation) University of California San Diego's Daniel Alspach Chair in dynamic systems and controls European Union Jean Monnet Chair at the New York University's School of Law David Mulvane Ehrsam and Edward Curtis Franklin Chair in chemistry at the University of Toronto's Faculty of Applied Science and Engineering (See Table 8.3)

(continued)

Table 8.2 (continued)

Causes	RY	RM	IY	IM	Best practices	Some case examples
(j) Branding: lack of brand or of experienced research team	x		x		(j1) Partner with complementary brands and write mediatic reports	Boston Consulting Group Perspectives MIT Technology Review—Opinno Harvard Business School Review—Opinno
					(j2) Review the processes of your communication unit: function map, communication processes in cascade and CRM of specialized media	University of Michigan Duke University's Corporate Relations Office
					(j3) Partner with visiting researchers or reward recognized faculty	IBM Faculty Awards
(k) Transfer: unsellable or incomprehensible products	x		x		(k1) Do periodic lectures to industry translating research results into qualified impact	French National Centre for Scientific Research Barcelona Supercomputing Center
(l) Transfer: disproportionate size of the research team		x			(l1) Adapt the size of the teams to the market needs	Several Spanish research centers
(m) Transfer: internal bureaucracy and politics		x		x	(m1) Identify the decision makers and identify their key performance indicators	Merck—Word Wide Licensing and Knowledge Management group
(n) Transfer: nonacceptance of generated research results	x		x		(n1) Define the delivery requirements or deadlines before starting, and presell your solution	Audi—Technical University of Munich
(o) Transfer: lack of public funding	x		x		(o1) Create a specific unit to apply for public funds and leverage external research incubators	Ateknea Max Plank IT Inkubator—Saarland University

(continued)

Table 8.2 (continued)

Causes	RY	RM	IY	IM	Best practices	Some case examples
All stages—Innovation ecosystem: catalyst prioritization						
(p) Agents—lack of understanding or mapping of the innovation ecosystem	x	x	x		(p1) Map the key stakeholders of your innovation ecosystem	Stanford University's Triple Helix research group
						Tel Aviv University
						National agency MATIMOP
					(p2) Identify the quality of the agents and the ecosystem	Apple's Research Center
(q) Agents—gaps in the innovation ecosystem (internally)	x		x		(q1) Map and connect with your organizational chart, influencers and advisors	Merck—Word Wide Licensing and Knowledge Management group
(r) Agents—gaps, little development or no proximity to the agents of the innovation ecosystem (externally)	x	x	x	x	(r1) Adapt your commercialization model to the characteristics of your innovation ecosystem	Sweden Chalmers University
						Cambridge University
						MIT Regional Entrepreneurship Acceleration Program
					(r2) Connect virtually with disperse agents	European Forum for Entrepreneurship Research
						Canadian Networks of Centres of Excellence
						Rolls-Royce Research—University Technology Centres
(s) Agents—no resources within your innovation ecosystem (internally) and difficulty in retaining talent	x	x	x	x	(s1) Crowdsource the areas of your value chain that are not in the core of your business model and specialize each function	Tel Aviv University's Ramot
						The Boston Consulting Group's Henderson Institute
						Accenture's Institute for High Performance
						PwC's Health Research Institute
					(s2) Capitalize on aging	IESE Business School
(t) Relationships—few interactions among agents	x		x		(t1) Move from academics to entrepreneurial academics	University of California's San Diego branch
						The Stockholm School of Entrepreneurship
					(t2) Recognize science entrepreneurs before they leave	Amazon
						Cambridge Science Park

Source Prepared by the author

8.3 A Case Example for Each Collaborative Business Model

See Table 8.3.

Table 8.3 A case example for each collaborative business model

Collaborative business model	Example
Short-term external contracting	UC Berkeley's California Center for Innovative Transportation
Medium-term external contracting	A.T. Kearney Global Business Policy Council
Long-term external contracting	Imperial College London Shell
Transfer pricing	GE Global Research Center
Freemium product/service	University of Southern California's Founder Central initiative
Research licensing	Max Planck Innovation
Technology transfer by public funding	NASA
Creation of a spin-off from the research center via external investment	Shire University of Melbourne's Fibrotech Therapeutics
Search model	Roche
Consultancy joint venture	Columbia Business School PwC's Strategy& (former Booz & Company)
Short-term marketing collaboration	National Science Foundation's Engineering Research Centers
Long-term marketing collaboration	Stanford University School of Humanities and Sciences

Source Prepared by the author

8.4 Research Methodology

This research project sets out to find answers to a main research question as well as five sequential and interconnected subresearch questions. The main query was: why do research centers fail to capture economic value from their knowledge assets and how can they improve without damaging research quality? In other (and more positive) words: how can research centers improve the commercialization of their discoveries?

The subquestions are related to the challenges, disconnections, symptoms, problems, and solutions. First, what do managing directors at research centers consider to be the main challenges to commercializing their internal discoveries? Second, in the commercialization process of a new discovery, what are the broken connections among the stages of the innovation funnel for commercializing a discovery within a research center? Third, what are the symptoms—understood as the two most common scenarios—of those broken connections? Fourth, what barriers are causing those broken connections? Finally, what is the best practice for solving each of the identified broken connections?

To answer these questions, the 4-year project followed a quantitative and qualitative analysis. Quantitatively, it used triangulation and data regression with different databases: a group of 3,881 research centers in 107 countries with data from 2009 to 2014, another group of 125 Spanish research centers and, finally, a group of some of the top players worldwide. The first group was measured according to 19 metrics, the second group according to 53 metrics and the last group with an open pattern (See Table 8.4).

Qualitatively, there was in-depth analysis of 35 research centers from four different countries, as well as 61 interviews—42 one-on-one and 19 with panels—involving politicians, consultants, investors, advisers, professors, executives, and researchers. Additionally, the institutions of 54 out of the 61 interviewees were visited and examined on-site. A review of 327 publications has been done.

However, I acknowledge four methodological limitations of this study: first, executives at research centers did not always want to discuss all of their problems for obvious reasons of confidentiality. As a consequence, it is possible they faced other obstacles but kept quiet about them. I tried to triangulate and complement those interviews with additional public data, supplementary interviews, and on-site visits to 28 research centers.

Second, a wider scope of the sample may increase understanding of the phenomena and the mechanisms for implementing the process. Nonetheless, among the sample, a group of top performers was chosen to increase the chances of aggregating the mechanisms in the same sample. The information was also expanded using academic and white papers on best practices.

Third, the analysis done comparing the size of research teams with total funding and indexed papers at research centers (See Fig. 5.16) is not an isolated experiment with independent and dependent variables but a gathering of indicators. Therefore, the results illustrate not cause–effect relation but just correlation. An additional experiment can be done ensuring the size of research team as an independent variable and keeping the other variables as dependent.

Fourth, the chosen sample of research centers includes different sectors. Each sector behaves differently and therefore the aggregation of data in the analysis of these groups may lead to incorrect interpretations. Nevertheless, I analyzed the differences among sectors using different metrics including economic, academic, and social impact to understand those dissimilarities.

8.5 Correlation Analysis Between Several Performance Metrics at Research Centers

See Table 8.4.

8.6 Author Biography

Josemaria Siota is a project leader at IESE Business School's Entrepreneurship and Innovation Center and the director of Simastec Consulting. His field of work focuses on strategic growth and innovation in corporations, universities, and government.

On this topic, he has published books, studies, and articles (e.g., for McGraw-Hill, PwC, Oliver Wyman, and *The European Financial Review*), spoken at conferences (e.g., Harvard Business School) and earned media mentions (e.g., *Forbes*).

Table 8.4 Correlation analysis between several performance metrics at research centers (N = 125).

	1	2	3	4	5	6	7	8	9	10	11	12	13	14	15	16	17	18	19	20	21	22	23	24	25	26	27	28	29	30	31	32	33	34	35
Research projects (#)																																			
1 Total	1.00	0.39	0.08	0.00	0.00	0.16	0.68	0.82	0.33	0.13	0.13	0.03	0.00	0.00	0.01	0.27	0.04	0.10	0.02	0.07	0.00	0.00	0.01	0.02	0.06	0.05	0.01	0.13	0.00	0.03	0.30	0.48	0.29	0.45	0.01
2 National calls		1.00	0.17	0.00	0.02	0.09	0.38	0.10	0.44	0.26	0.25	0.06	0.02	0.01	0.13	0.06	0.07	0.10	0.03	0.13	0.05	0.08	0.07	0.13	0.07	0.08	0.09	0.10	0.00	0.08	0.43	0.61	0.61	0.62	0.02
3 European calls			1.00	0.00	0.00	0.00	0.12	0.00	0.47	0.07	0.46	0.00	0.01	0.00	0.07	0.13	0.09	0.02	0.00	0.21	0.01	0.02	0.01	0.05	0.00	0.16	0.17	0.19	0.01	0.03	0.24	0.20	0.14	0.23	0.01
4 Local calls				1.00	0.03	0.00	0.00	0.02	0.01	0.02	0.00	0.17	0.02	0.00	0.00	0.00	0.01	0.00	0.00	0.02	0.01	0.01	0.01	0.01	0.13	0.03	0.01	0.02	0.03	0.00	0.00	0.00	0.00	0.00	0.23
5 Other local calls					1.00	1.00	0.00	0.02	0.00	0.00	0.00	0.01	0.00	0.66	0.00	0.01	0.00	0.04	0.00	0.03	0.00	0.00	0.00	0.01	0.05	0.00	0.01	0.14	0.09	0.01	0.01	0.02	0.03	0.02	0.12
6 Other international calls						1.00	0.23	0.04	0.26	0.02	0.27	0.00	0.00	0.00	0.18	0.25	0.00	0.03	0.01	0.06	0.00	0.00	0.00	0.02	0.03	0.01	0.02	0.15	0.00	0.03	0.23	0.27	0.16	0.28	0.02
7 Public funds							1.00	0.38	0.37	0.14	0.23	0.01	0.01	0.02	0.00	0.34	0.04	0.12	0.01	0.08	0.01	0.01	0.01	0.03	0.05	0.00	0.01	0.12	0.00	0.03	0.46	0.58	0.28	0.57	0.05
8 Private funds								1.00	0.09	0.03	0.03	0.01	0.01	0.01	0.01	0.17	0.02	0.04	0.01	0.01	0.00	0.00	0.00	0.00	0.01	0.01	0.02	0.03	0.00	0.00	0.08	0.18	0.09	0.15	0.01
Research funding (€)																																			
9 Total									1.00	0.42	0.58	0.10	0.00	0.00	0.05	0.25	0.10	0.14	0.01	0.19	0.00	0.00	0.00	0.03	0.03	0.07	0.15	0.19	0.01	0.06	0.48	0.58	0.24	0.57	0.00
10 National calls										1.00	0.01	0.20	0.00	0.00	0.00	0.10	0.07	0.06	0.00	0.13	0.01	0.01	0.01	0.01	0.02	0.03	0.08	0.04	0.04	0.06	0.33	0.30	0.10	0.32	0.00
11 European calls											1.00	0.00	0.00	0.01	0.13	0.02	0.05	0.07	0.00	0.09	0.01	0.03	0.02	0.03	0.00	0.02	0.02	0.16	0.00	0.01	0.21	0.32	0.14	0.30	0.00
12 Local calls												1.00	0.00	0.00	0.00	0.00	0.00	0.02	0.00	0.01	0.00	0.00	0.00	0.01	0.03	0.02	0.05	0.01	0.03	0.01	0.06	0.05	0.03	0.05	0.00
13 Other local calls													1.00	0.00	0.00	0.00	0.01	0.01	0.00	0.03	0.00	0.00	0.01	0.00	0.00	0.01	0.00	0.08	0.03	0.00	0.01	0.02	0.04	0.02	0.12
14 Other international calls														1.00	0.02	0.00	0.00	0.00	0.00	0.00	0.00	0.00	0.00	0.00	0.03	0.00	0.00	0.01	0.01	0.01	0.01	0.01	0.01	0.01	0.00
15 Public funds															1.00	0.03	0.07	0.00	0.00	0.05	0.00	0.00	0.00	0.01	0.00	0.00	0.00	0.03	0.02	0.01	0.16	0.28	0.13	0.25	0.05
16 Private funds																1.00	0.07	0.05	0.00	0.03	0.00	0.00	0.00	0.00	0.01	0.04	0.22	0.02	0.06	0.02	0.11	0.13	0.06	0.13	0.00
Technology transfer																																			
17 International patents																	1.00	0.25	0.00	0.01	0.02	0.01	0.01	0.00	0.01	0.00	0.02	0.07	0.02	0.00	0.05	0.06	0.01	0.06	0.00
18 Licensing contracts																		1.00	0.00	0.15	0.01	0.01	0.01	0.00	0.19	0.00	0.04	0.14	0.02	0.00	0.13	0.11	0.03	0.12	0.00
19 Spin-offs																			1.00	0.00	0.00	0.00	0.00	0.00	0.00	0.00	0.00	0.00	0.00	0.00	0.03	0.02	0.01	0.02	0.00
Publications																																			
20 Refereed articles																				1.00	0.01	0.00	0.00	0.02	0.10	0.12	0.13	0.14	0.06	0.03	0.26	0.17	0.06	0.21	0.04
21 Nonrefereed articles																					1.00	0.65	0.71	0.50	0.00	0.16	0.01	0.00	0.02	0.33	0.03	0.00	0.08	0.01	0.02
22 Books																						1.00	0.83	0.54	0.00	0.21	0.00	0.00	0.03	0.33	0.00	0.09	0.09	0.01	0.03
23 Book chapters																							1.00	0.56	0.00	0.20	0.00	0.00	0.03	0.41	0.02	0.00	0.07	0.01	0.03
Congresses																																			
24 National communications																								1.00	0.05	0.23	0.01	0.03	0.06	0.31	0.07	0.03	0.09	0.05	0.00
25 National posters																									1.00	0.01	0.16	0.19	0.03	0.00	0.11	0.06	0.03	0.08	0.04
26 International communications																										1.00	0.13	0.02	0.11	0.32	0.07	0.04	0.07	0.06	0.01
27 International posters																											1.00	0.07	0.06	0.10	0.23	0.13	0.10	0.18	0.00
Training																																			
28 Ph.D. thesis																												1.00	0.08	0.04	0.07	0.03	0.09	0.05	0.00
29 Postgraduate courses																													1.00	0.09	0.11	0.06	0.03	0.08	0.04
30 Events																														1.00	0.10	0.05	0.10	0.18	0.00
Employees																																			
31 Academics																															1.00	0.71	0.37	0.85	0.05
32 Assistants																																1.00	0.56	0.96	0.04
33 Management																																	1.00	0.60	0.11
34 Total internal																																		1.00	0.05
35 External																																			1.00

Source Prepared by the author with data from the Spanish National Research Council. *Datos* (2013) [12]. The selection of indicators was chosen based on a literature review that included articles such as Bozeman, B., Rimes, H. & Youtie, J. The evolving state of the art in technology transfer research: Revisiting the contingent effectiveness model. *Research Policy* 44, 34–49 (2015). [13]. *Note* The table includes the quadratic correlation coefficients of the 35 out of 53 analyzed metrics

Moreover, he has coached aspiring executives from HBS, MIT-Sloan, Stanford-GSB, Wharton, LBS, and INSEAD. He has also cofounded five start-ups and cocreated several initiatives at IESE in different countries.

Previously, he was a consultant at Deloitte and earned his two master's degrees from the Polytechnic University of Catalonia while also attending courses at IESE.

References

1. George, G., Zahra, S. A. & Wood, D. R. The effects of business-university alliances on innovative output and financial performance: a study of publicly traded biotechnology companies. *Journal of Business Venturing* **17**, 577–609 (2002).
2. Azároff, L. V. Industry–University collaboration: how to make it work. *Research Management* **25**, 31–34 (1982).
3. Bonaccorsi, A. & Piccaluga, A. A theoretical framework for the evaluation of university–industry relationships. *R & D Management* **24**, 229–247 (1994).
4. Hall, B. H. *et al.* Universities as research partners. *The Review of Economics and Statistics* **85**, 485–491 (2003).
5. Hsu, D. W. L., Shen, Y. C., Yuan, B. J. C. & Chou, C. J. Toward successful commercialization of university technology: performance drivers of university technology transfer in Taiwan. *Technological Forecasting and Social Change* **92**, 25–39 (2015).
6. Hurmelina, P. *Seminar on Innovation: Motivations and Barriers Related to University–Industry Collaboration-Appropriability and the Principle of Publicity* (Haas, 2004).
7. Shane, S. University technology transfer to entrepreneurial companies. *Journal of Business Venturing* **17**, 537–552 (2002).
8. Azagra-Caro, J. M., Archontakis, F., Gutiérrez-Gracia, A. & Fernández-de-Lucio, I. Faculty support for the objectives of university–industry relations versus degree of R & D cooperation: the importance of regional absorptive capacity. *Research Policy* **35**, 37–55 (2006).
9. Meyer-Krahmer, F. & Schmoch, U. Science-based technologies: university–industry interactions in four fields. *Research Policy* **27**, 835–851 (1998).
10. Gulbrandsen, M. & Smeby, J. C. Industry funding and university professors' research performance. *Research Policy* **34**, 932–950 (2005).

11. Laukkanen, M. Exploring academic entrepreneurship: drivers and tensions of university-based business. *Journal of Small Business and Enterprise Development* **10**, 372–382 (2003).
12. Spanish National Research Council. *Datos* (2013).
13. Bozeman, B., Rimes, H. & Youtie, J. The evolving state-of-the-art in technology transfer research: revisiting the contingent effectiveness model. *Research Policy* **44**, 34–49 (2015).

Index

A

Academic
   engagement 8, 130
   impact 100
   intrapreneurship 138
   rigor 4, 31, 38, 46
Accenture 6, 134
Adobe 85
Aging capitalization 134, 135
Amazon 139
Ateknea 106
A.T. Kearney 77
Auckland savings bank 49
Audi 104, 105

B

Barcelona super computing 98
Barcelona Technology Transfer Group (BTTG) 58
BNP paribas 49
Boston Consulting Group 99, 134
Brazil science park 39
Broken innovation 16, 17, 20, 21, 28, 29, 44, 45, 69, 117, 144
Buckman laboratories 59
Business angel 122, 131
Business model 6, 44, 45, 49, 68, 69, 72, 73, 81, 83, 85, 93, 124, 131, 134, 139, 144

C

California 48, 75, 81, 87
Cambridge 131, 140
   science park 140
   university 36, 131
Center for genomic regulation 50, 60
China 116, 124, 130
   Industrial and Commercial Bank 49
Conceptual selling 129
Cornell Tech 38, 59, 68
Crowdsource 133, 134

Customer journey map 44, 49
Customer relationship management (CRM) 96, 97

D

Deutsche Bank Research 77
Deutsche Telekom laboratories 60
Discovery 14, 16, 48, 54, 57, 98, 99

E

Economic 3, 5, 8, 15, 28, 30, 34, 62, 77, 104
　impact 32, 39, 100
　sustainability 3, 18, 31, 38, 144
Economist Group, The 77
Entrepreneurial 87, 116, 135, 137, 139
　scientist 136, 137
　university 136, 140
European Forum for Entrepreneurship Research(EFER) 132

F

Ford 85
Fraunhofer 82, 93, 144

G

General electric 32, 79
Germany 57, 70, 77, 80, 82, 86, 106, 107
Global innovation index 124, 125
Google 37, 75, 77, 80

H

Harvard 144
　Business Review 95
　Business School 81, 95, 138
　Innovation Lab 58
Health Canada 37
Hewlett-Packard 74
　laboratories 74
Horizon2020 72, 105, 107
Humboldt innovation 57

I

IBM 60, 85, 98
　faculty awards 98
　research 98
IDEO 48
Impact factor 54
Industry
　collaboration 3, 16, 59, 70, 72, 98, 100
　engagement 7, 59
Innovation 2, 4, 6, 14, 15, 20, 21
　ecosystem 16, 17, 20, 32, 73, 83, 88, 117, 118, 121, 124, 127, 130, 144
　funnel 15, 16, 18, 20, 21, 44, 68, 118, 121
Innovation, lab 49
INSEAD 124, 126
Intel 49, 85
Internal contracting 74, 79
International advisory board 39
Invention 46, 87
Investor in residence 68
Isis Innovation 87
Israel 116

J

Jigsaw 77

Index

K
Key performance indicators. *See* KPIs
Knowledge 4, 5, 36, 38, 39
    asset 7, 8, 15, 44, 50, 69, 136
    push 21, 52
    sharing 29, 35, 37, 135

L
Lean, research 55, 56
Linked innovation 14, 21, 28, 45, 46, 69, 144
Long-term 53, 77, 97
    contracting 73, 77, 78
    marketing collaboration 74, 92
LSE business review 49

M
Market
    need 99
    pull 20, 21, 52
Massachusetts Institute of Technology. *See* MIT
Max Planck innovation 81, 106
McKinsey 32
    Global Institute. *See* MGI
McKinsey Global Institute (MGI) 40
Mentoring 58, 61, 62, 91
MIT 28, 61, 99, 130, 131
    Deshpande Center 61
    Design Lab 55
    research team 61
    Sloan Management Review 157
    Technology Review 99

N
National Aeronautics and Space Administration. *See* NASA
Nokia 75
Norway 72

O
Opinno 95
Organization for Economic Cooperation and Development. *See* OECD
Oxford University 87

P
Paper prototyping 44, 49
Performance metrics 3, 16, 18, 28, 32, 33, 45, 46, 59, 61, 104, 144
Petropolis Technopole 39
Pittsburgh 39
PwC 77, 88, 125, 134
    Health Research Institute 134

R
Ramot 116, 122
Regional Entrepreneurship Acceleration Program. *See* REAP
Research
    center 22, 31, 33, 34, 38, 49, 56, 60, 61, 69, 70, 73, 75, 77, 79, 83, 85, 87–89, 93, 95, 98, 100, 102, 117, 118, 124, 126, 129–131, 137, 139, 144
    collaboration 69, 72, 74
    funding 107
    incentives 140
    team 32, 35, 37, 48, 56, 57, 60, 62, 68–70, 100, 130
Resource

allocation 5
scarcity 134
Roche 14, 36, 88, 144
Rolls-Royce 133
research 133
Root capital 56
Royal bank 49

S

Samsung Economic Research Institute 77
SAP 58, 81, 85
Next-Gen 58
Scimago 124
Shell 79
Short-term
external contracting 74
marketing collaboration 74, 89
Small and medium enterprises(SMEs) 70
Social impact 28, 34
South Africa 126, 127
Spain 95, 100, 124, 125, 131
Spin-off 33, 38, 57, 68, 74, 83, 85, 86, 131, 137
Stanford University 44, 48, 92, 121, 131
Triple Helix research group 135
Stockholm School of Entrepreneurship 136

Sweden 124, 127
Chalmers University 131

T

Technical University of Munich. *See* TUM
Technology transfer 57, 74, 82, 86, 98, 118, 121
Technology transfer office 50, 57, 58, 137
Tel Aviv University 116, 122
Triple Helix model 121, 122
Tufts Fletcher School 55

U

UC Berkeley 75
United Kingdom 70, 77, 80, 124, 130
United States 4, 36, 49, 60, 70, 75, 89, 127, 130, 131, 133, 137
University of California San Diego 92, 136
University of Melbourne 39, 86
University of Toronto 93

W

Wells Fargo 49
World Bank 36

The manufacturer's authorised representative in the EU is Springer Nature Customer Service Centre GmbH, Europaplatz 3, 69115 Heidelberg, Germany. If you have any concerns regarding our products, please contact ProductSafety@springernature.com

Printed and bound by CPI Group (UK) Ltd, Croydon, CR0 4YY

23/03/2026

02076402-0015